Watching Nature

10/10/02

Happy Birthday Jerry.
Best wishes for many,
many more.
Happy birding!
    Love,
    Joe & Brenda

Watching

A MID-ATLANTIC

NATURAL HISTORY

MARK S. GARLAND

WITH ART BY JOHN ANDERTON

SMITHSONIAN INSTITUTION PRESS
WASHINGTON AND LONDON

Editor: Catherine Puckett Haecker
Production editor: Jack Kirshbaum
Designer: Janice Wheeler
Typesetting and maps: Blue Heron Typesetters, Inc.

Library of Congress Cataloging-in-Publication Data
Garland, Mark S.
    Watching nature : a Mid-Atlantic natural history / Mark S. Garland
  : with art by John Anderton.
      p.   cm.
    Includes bibliographical references and index.
    ISBN 1-56098-742-1 (paper)
    1. Natural history—Middle Atlantic States.   2. Natural history—Virginia.
  3. Natural history—West Virginia.   I. Title.
  QH104.5.M45G37   1997
  508.75—dc20   97-846

British Library Cataloging-in-Publication data available

Manufactured in the United States of America
02 01 00 99 98 97   5 4 3 2

The paper used in this publication meets the minimum requirements of the American National Standard for Permanence of Paper for Printed Library Materials Z39.48-1984.

To the memory of
ROBERT COLEMAN GARLAND,
my father and my best friend.

Contents

Preface   ix

I. ALLEGHENY TO ATLANTIC: AN OVERVIEW OF MID-ATLANTIC
NATURAL HISTORY   I

The Seasons   41

2. SPRING: MARCH THROUGH MAY   43

3. SUMMER: JUNE THROUGH AUGUST   58

4. AUTUMN: SEPTEMBER THROUGH NOVEMBER   69

5. WINTER: DECEMBER THROUGH FEBRUARY   85

*Places to Explore: Six Journeys through the
Mid-Atlantic*   101

6. WEST VIRGINIA HIGHLANDS: NATURE DOWN THE
HIGHLAND TRACE   103

7. THE POTOMAC: THE NATION'S RIVER   118

8. RIDGE AND VALLEY REGION: APPALACHIAN TRAIL
COUNTRY   141

9. THE BAY: EXPLORING LANDS AROUND THE CHESAPEAKE   157

10. MID-ATLANTIC COAST   178

11. CLOSE TO HOME: NATURE NEAR BALTIMORE AND
    WASHINGTON, D.C.   195

EPILOGUE   225

Appendix 1. A Beginning Naturalist's Reference Library   227

*Appendix 2. Conservation Organizations, Nature Centers, Parks,
Wildlife Refuges, and Related Resources*   229

*Appendix 3. Common and Scientific Names of Plants and
Animals*   238

References   249

Index   257

Preface

This book is a love story. It's an expression of my love of exploring nature in the region I call home, the mountains from West Virginia to Pennsylvania, the coast from southern New Jersey through Virginia, and many places in-between. It's a book about those places that I love—mountains, rivers, beaches, marshes, forests, and fields. It's also a book about what I love to explore—trees and flowers, birds and butterflies, salamanders and rocks. I hope, too, that there is some practical value to the book, that it can help lead others to fascinating places and wonderful discoveries. More than that, I hope this book expresses the wonder I find every day in this amazing world, and that it inspires the reader to explore, discover, love, and protect our natural environment.

You'll find no cookbook for nature study here, no explicit directions to particular places with long lists of what you'll see. Others have done this, and I've listed many of their excellent books in the references. But I invite you to join me as I ramble through these pages, telling stories and sharing favorite finds from many of our region's premier natural areas. It's my hope that you'll find here a full sense of place—the big picture. I'll feel successful if I nurture some basic understandings of the complex assemblages of rocks and dirt and plants and animals that are nature in the mid-Atlantic region.

I've had a lot of help putting this book together. Thanks to all who have spent time in the field with me. I've had the good fortune to take field trips to many parts of the mid-Atlantic with literally thousands of people over the last 25 years. I can't express my gratitude to them all, but I'd like to single out some people whose teaching, encouragement, inspiration, and companionship in the field have been especially helpful:

Kim Bierly, John Bjerke, Rich Bray, Ronald L. Canter, Paige Cunningham, Kathy Dale, Janet Dierker, Margaret Donnald, Morrill Donnald, Miles Drake, Nathan Erwin, Tom Feild, Neal Fitzpatrick, Cris Fleming, Vagn Flyger, Harry C. Garland, Mary T. Garland, Robert C. Garland, Robert T. Garland, Rob Gibbs, Denise Gibbs, Laura Gaudette Green, Helen Kavanagh, Ivan Klein, Roger Klinger, Tamar Krichevsky, Larry Lang, Helen G. Mackintosh, Stephanie Mason, Donald H. Messersmith, Karyn Molines, Gary Mozel, William L. Murphy, Lola Oberman, Richard Orr, Joseph Patt, Gary Pendleton, Robert Michael Pyle, Tim Ray, Joel Rhymer, Michael Rosen, Jack Schultz, Lora Seraphin, Marti Seraphin, Michael E. Seraphin, Michael R. Seraphin, Stanwyn G. Shetler, Robert C. Simpson, Richard H. Smith, Darryl Speicher, David W. Sturman, Mark Swick, William H. Triplett, John Trott, Keith Van Ness, Natalie Venneman, Leo Weigant, Jan Westervelt, Tony White, Hal Wierenga, Claudia Wilds, Erika Wilson, and Emmett L. Wright.

I give special thanks to Ken Nicholls, former executive director of the Audubon Naturalist Society, for encouraging me to take nature writing seriously. I also thank members of the editorial staff of the *Audubon Naturalist News*, particularly Kathryn Karsten Rushing, Leslie D. Cronin, and Barbara Tufty, who nurtured me when I was a beginning writer. Portions of this book are modified from articles that originally appeared in the *Audubon Naturalist News*, a publication of the Audubon Naturalist Society. Thanks to those who made comments on early drafts of this manuscript, Kent and Marcia Minichiello, and anonymous reviewers—this is a far better book thanks to their valuable suggestions. Thanks to John Anderton for his amazing illustrations; I hope the words are worthy of his fine art. The staff at Smithsonian Institution Press has earned my deep gratitude and respect. Thanks to Peter Cannell for his vision and for his confidence in me. Thanks also to Jack Kirshbaum, Janice Wheeler, Catherine Puckett Haecker, Ken Sabol, Kate Gibbs, and the rest of that fine staff.

Because I have a tendency to be absentminded, it's likely that I've forgotten a few important people; I hope they will forgive me. In spite of all this excellent assistance, there certainly could be some errors in the text, and these, of course, are solely the responsibility of the author.

1

Allegheny to Atlantic: An Overview of Mid-Atlantic Natural History

Nature is all around us. It is the landscape on which we live, the climate that brings us weather, the plants and animals that share the planet's surface with us. Those plants produce the oxygen we breathe and support the life of all animals, including ourselves, either directly or indirectly. We are part of nature, and our lives depend on nature every moment of every day. Our busy lives make it easy to forget nature, however, especially if we live in urban and suburban areas. This book is for people who don't want to forget.

The region covered by this book is an irregularly shaped area surrounding Washington, D.C. and Baltimore. These major cities lie in the midst of the

Beaver and pond

great megalopolis of the eastern U.S., an extensively developed and densely populated region that stretches along much of the East Coast. At first glance this region may seem like a poor one for nature study, yet it includes a surprising number of highly diverse natural areas to explore, including elements of southeastern ecosystems, northern boreal environments, and a variety of transitional habitats. Areas less than 150 miles away from Washington, D.C. include varied highlands of the Allegheny Mountains and the Blue Ridge, surprisingly pristine river systems from the Potomac to the Pocomoke, rich coastal and marine environments of the Atlantic Ocean and the Chesapeake Bay, and southern forests with subtropical affinities. The region's central location in eastern North America, combined with the cooling effect of altitude and the warming influence of the ocean, enables all but the most extreme habitats of the eastern half of the continent to exist in one place or another.

Naturalists living in the Washington-Baltimore region have many plants and animals at hand to learn about and to observe, and have many habitats to explore. That's what I've been doing for much of my life, and with this book I hope to provide a general overview of the natural history of this region, loosely defined as that portion of the mid-Atlantic within a 4-hour drive of the Washington, D.C. and Baltimore metropolitan centers. Along the way I'll tell a few stories of my own explorations. This chapter provides a general framework for such a study. Subsequent chapters embellish this introduction, focusing on some seasonal natural history highlights and looking at specific natural areas of interest. My intent is to provide a starting point for an understanding and appreciation of nature in this region.

BASIC MID-ATLANTIC ECOLOGY

The mid-Atlantic region is neatly divided geologically into these four provinces: the Allegheny Plateau, Ridge and Valley, Piedmont, and Coastal Plain. These fundamental subdivisions are used not only to define area and geological origins, but also to describe plant and animal communities, for each of the four provinces is home to ecosystems whose characteristics are distinctly different from each other.

The ecological communities found in any given region are influenced by a combination of climatic factors and features of the landscape, both now and in the past. Landscape, climate, and history determine a region's plant communities, which, in turn, influence the wildlife found in each place. All these factors influence past and present human settlement and development of the region, which profoundly affect all our ecosystems, on local and global scales.

I use the following basic framework to develop an understanding of any region's natural history. First, examine the landscape and climate. Next, study the plant and animal communities. Finally, factor in the influence on the landscape wrought by people—both destructive and protective activities. Let's follow this model for the mid-Atlantic.

GEOLOGY

Roughly 250 million years ago, during the late Paleozoic era and just before the rise of dinosaurs, the combined land mass of Europe and Africa pushed against that of the Americas. This pressure, which continued for millions of years, caused an enormous uplift of the eastern edge of North America; the entire landscape was pushed tens of thousands of feet upward. Close to the impact zone, the existing sedimentary and igneous rocks were twisted and folded, broken along large planes called faults, and physically changed into metamorphic rocks by tremendous subterranean heat and pressure. Farther west less pressure resulted in gentle folds, fewer faults, and less metamorphism.

What goes up must come eventually down. As the land began to rise, erosive forces increased. The mountains wore down even as they rose, though not as quickly. The uplift of eastern North America, known as the Appalachian orogeny, ended more than 200 million years ago, but erosion continues to this day. The present landscape of eastern North America reflects those uplift periods of long ago and their continuous erosion.

Because the westernmost uplifted areas were relatively removed from the pressure front, they showed less folding and soon eroded down to highly resistant layers. These layers slowed the erosive rate dramatically, leaving an elevated landscape that exists today as the Allegheny Plateau. The westernmost areas considered in this book include the West Virginia Highlands and Garrett County, Maryland, both areas on the Allegheny Plateau.

East of the Allegheny Plateau lies the Ridge and Valley province, a region of long, parallel ridges separated by broad, low valleys. Here the pressures of mountain building were more severe, resulting in greater twisting

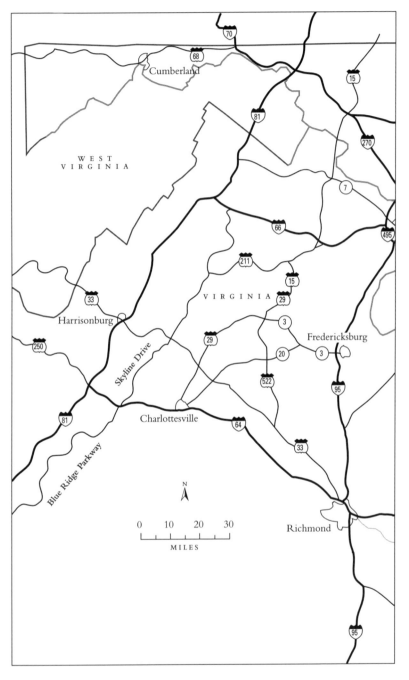

Map 1. Mid-Atlantic Region

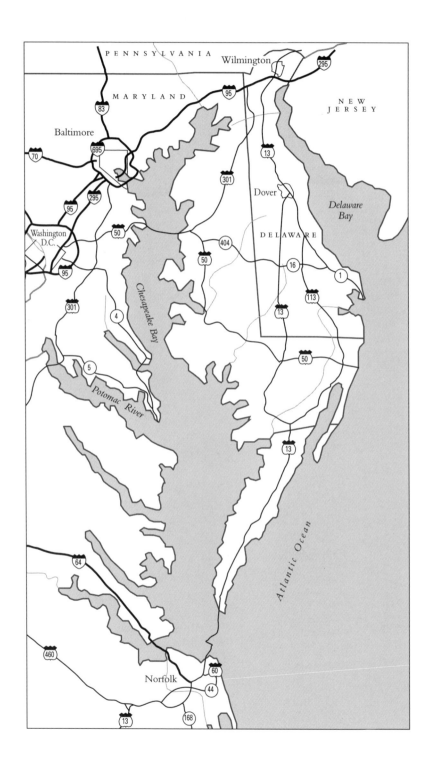

and faulting of the bedrock. Resistant rock layers occur here as they do on the Allegheny Plateau, but they are heavily fractured. Because fractures are subject to much more rapid erosion than are unbroken rock layers, erosion has been far more substantial in the Ridge and Valley region than on the Allegheny Plateau. In contrast to the broad highlands and narrow, elevated valleys of the Allegheny Plateau, the Ridge and Valley features narrow ridges, most considerably lower in elevation than the highlands of the plateau, and broad, low valleys, many less than 1,000 feet in elevation. The Great Valley, which is also called the Shenandoah Valley in Virginia and the Hagerstown Valley in Maryland, is the broadest of these lowlands. The Blue Ridge, home to Shenandoah National Park and a section of the Appalachian Trail, forms the eastern edge of the Great Valley and, in most places, the eastern edge of the Ridge and Valley province. Most of the mountains of Virginia discussed in this book lie in the Ridge and Valley province, as do areas in eastern West Virginia, most of western Maryland, and much of central Pennsylvania.

The next region to the east is the Piedmont province, the term derived from French words meaning *foothill*. Here the landscape is lower and gently rolling with a few prominent ridges or isolated peaks that are called monadnocks. This region felt the full force of the Paleozoic collision of continents, when its rocks were twisted, gnarled, faulted, and metamorphosed. Erosive forces have taken the heights of the Piedmont almost down to sea level, with only the most erosion-resistant rocks showing significant relief, as at Sugarloaf Mountain in Maryland and Bull Run Mountain in Virginia.

The eastern edge of the Piedmont is called the Fall Line. It is at this geologic boundary between the resistant rock of the Piedmont and the softer sediments of the Coastal Plain that creeks and rivers tumble over their last falls and reach elevations within a few feet of sea level. It's easy to draw the Fall Line on a political map; most major mid-Atlantic cities developed on the Fall Line, including Richmond, Fredericksburg, Washington, D.C., and Baltimore.

All of the regions listed thus far have been eroded landscapes, where Earth's surface is primarily expressed as raised bedrock minus material that has been eroded away. East of the Fall Line lies the Coastal Plain province, which is basically a depositional landscape where bedrock occurs below sea level. The land that we see at the surface in the Coastal Plain province is material eroded from the highlands to the west and deposited here. A lot of material can wash down from the mountains in more than 200 million years! The entire landscape from the Fall Line to the Atlantic coast and, to some

extent, to the edge of the Continental Shelf, 50 or more miles ⌐
built up from material eroded down from the highland regions to the west.
Along many parts of the Atlantic coast these sediments are more than 2,000
feet thick.

CLIMATE

The mid-Atlantic region is part of the warm, moist-temperate climate zone.
Significant variation in precipitation exists in different portions of the region,
though all areas have enough rainfall for forests to exist as the climax vegetation. (In ecology, *climax* refers to the ecosystem that eventually develops in
the absence of disturbance.) Instead of pronounced wet and dry seasons, precipitation is fairly equally spread throughout the year. Temperature variations
are more extreme, with highland areas significantly colder than coastal areas
in every season. All parts of the mid-Atlantic see subfreezing temperatures
every winter.

Microclimates and their effects can be observed in many places. Microclimates exist in places where a small area's annual temperature or rainfall differ
from surrounding areas because of unusual topographic features. Some areas
with interesting microclimatic variations are noted later in this book, but
here I look only at the general trends for each region.

The Allegheny Plateau is the coldest and wettest province of the four. Its
colder temperatures are a result of altitude and distance from the sea. Much
of the Allegheny Plateau is above 3,000 feet (900 meters) in elevation, and
higher elevations are cooler than lower ones. When all other factors are
equal, a rise of 1,000 feet (300 meters) corresponds to a temperature drop of
about 3.5 degrees Fahrenheit. The Atlantic Ocean off the coast of the eastern United States is relatively warm because Gulf Stream currents move subtropical waters northward along the coast. The land closest to these warm
ocean waters benefits from this moderating influence on its climate.

Much of our region's precipitation is borne on prevailing westerly winds.
In all seasons, but most frequently during winter, moisture-laden low pressure systems regularly move across the continent. These air masses are forced
to rise as they reach the Allegheny Plateau mountain system. As the air rises
and cools, its moisture-holding capacity decreases, condensation occurs, and
it rains or snows. Although Washington, D.C. averages about 40 inches (100
centimeters) of precipitation per year, the average is over 50 inches (127 centimeters) at weather stations on the Allegheny Plateau.

The Ridge and Valley province is a little warmer and a lot drier. It is warmer because the elevations are lower, below 1,000 feet (300 meters) in most of the valleys. Some ridges reach 4,000 feet (1,200 meters) or more, but their average height is only about 2,000 feet (600 meters). This region is drier because much of the moisture coming across the continent has already been squeezed out by the Allegheny Plateau, which lies to the west. Annual precipitation in the Ridge and Valley ranges from 25 to 35 inches (63 to 89 centimeters), although the ridges are generally a bit colder and wetter than the valleys. This is most obvious in winter, when snow depth increases with altitude.

The overall lower elevations of the Piedmont result in warmer temperatures than in either of the mountainous provinces, but the annual precipitation—roughly 40 inches (100 centimeters) per year—is intermediate between that of the Allegheny Plateau and that of the Ridge and Valley. Although the Piedmont lies in the rain shadow of the Allegheny Plateau, this province and the Coastal Plain province usually pick up more summer precipitation than the mountain provinces—this is because summer weather systems bring hot, moist air northeastward from the Gulf of Mexico. This summer pattern brings rain to the entire region, often in the form of intense thunderstorms. Low pressure storm systems in summer and winter will sometimes stall off the Atlantic coast, blowing moist air inland from the ocean on winds blowing from the east or northeast. Often called *Nor'easters*, these storms can bring intense rainfall to the Coastal Plain, moderate rain to the Piedmont, and very little rain to the mountain zones farther west.

The Coastal Plain is our warmest region, thanks in part to the moderating influence of the Atlantic Ocean on winter temperatures. Precipitation in the Coastal Plain is similar to that of the Piedmont, averaging about 40 inches (100 centimeters) per year. Dramatic yearly variations can occur, resulting in one or two significant droughts per decade. Temperatures increase progressively to the south, as in all mid-Atlantic zones, with the southeastern corner of Virginia sometimes defined as subtropical, though only by the most liberal definition.

PLANT COMMUNITIES

In the absence of environmental disturbance, deciduous forest dominates most of our region's upland habitats; however, every environment in the eastern United States shows the effects of human disturbance. Virtually all our forests were cleared by the end of the nineteenth century. Many areas reverted to woodland in the twentieth century after intensive agricultural use left them unfertile, yet these forests surely don't support the richness of plants and animals of their precursors. Some species, such as the passenger pigeon and the Carolina parakeet, are gone forever. Our forests are recovering from past disturbances, and few could be called mature. Most mid-Atlantic forests have few trees older than 60 or 70 years.

Plant communities in every area reflect the potential climax vegetation combined with its stage of recovery from disturbance. Environmental disturbance can be caused by human activity or by natural processes. Logging, land clearing, development, overgrazing, and the introduction of nonnative species are a few of the human-caused environmental disturbances affecting plant communities. Natural disturbances include wildfires, floods, and storms.

When forest is removed from an area, a process called ecological succession takes place. In the first year or two after disturbance, sun-loving plants such as grasses, asters, goldenrods, and other herbaceous species thrive. The resulting habitat can be called either meadow or field. After a few years of growth, these herbs and the soil they help build allow shrubs and trees tolerant of sun and heat to sprout. These woody plants are called pioneer species and include Virginia pine, loblolly pine (on the Coastal Plain), tuliptree, cherry, black locust, and Virginia juniper (which is also called red cedar).

As the pioneer trees grow and create shade, the soil is augmented by the decay of herbaceous plants and the fallen leaves of the trees and shrubs. Once shaded, the soil is richer in organic material and holds moisture even longer, permitting the growth of yet another plant community, this one dominated by shade-tolerant trees. In most of our region these forests are composed of deciduous trees including various oaks, hickories, maples, American beech, birch, ash, black walnut, and basswood. The exception occurs in our highest mountains, where red spruce trees replace the pioneers. A forest of shade-tolerant trees is self-perpetuating, continuing to exist until a new environmental disturbance occurs. This final stage in the process of succession is called the climax. Plant communities at almost every location in the eastern United States are at some transitional (or subclimax) stage of the process of succession.

The Allegheny Plateau, our coolest region, harbors many plant communities with northern affinities. Most widespread is the northern hardwood forest community: rich, well-watered woodlands where sugar maple, basswood, yellow birch, black birch, American beech, and black cherry are among the dominant species. A few shrubs, herbs, and ferns are widespread in the understory of this region. The floor of this forest is fairly open in drier places, in contrast to the thick tangle of rhododendron or mountain laurel that grows in wetter places.

It's even colder on the highest points of the plateau, and in these areas an even more northern plant community thrives, that of the Canadian boreal forest. Red spruce is the dominant tree, with a mixture of the hardwoods just listed and, in places, hemlock and balsam fir trees. Wetlands are fairly common high on the plateau, thanks to impermeable rock layers near the surface and the activities of beavers. Balsam fir is most often found in these wetlands; from Maryland north they are joined by tamaracks. A number of northern plants and animals reach the southern edge of their ranges in the Allegheny Plateau mountain bogs and swamps.

The warmer, drier Ridge and Valley province supports hardwood forests dominated by a variety of oaks. American chestnut was formerly codominant here, but an introduced fungal blight has devastated this tree, changing its niche from canopy codominant to occasional shrub. Hickories and a few other hardwoods are typically mixed with the oaks, and tuliptrees are plentiful in younger forest stands. Fertile floodplain soils in the Ridge and Valley retain moisture far longer than the surrounding areas, resulting in a rich forest. Silver maple, box elder, slippery elm, sycamore, and black walnut are a few floodplain species. Spring wildflowers are abundant along these rivers, though farming and grazing have eliminated many of these natural gardens.

Belts of sedimentary limestone, sandstone, and shale in the Ridge and Valley support distinct vegetative communities. When limestone breaks down, it results in soil with high pH; certain plants, such as the walking fern, thrive only on these limestone soils. Sandstone breaks down to sandy soils; where sandy soil occurs on this province's dry ridges, only plants adapted to withstand severe drying can survive. Shale also breaks down into soil that retains moisture poorly. Shale slopes that face south absorb much heat on sunny summer days; ground temperature can exceed 150 degrees Fahrenheit. Few plants can tolerate such extreme temperatures. Some that can survive on these shale barrens are endemics, plants that grow in this habitat and no place else on Earth. Shale barren endemics include shale onion, Kates Mountain clover, whitehaired leatherflower, shale bindweed, and swordleaf phlox.

Piedmont forests are not as dry as those of the Ridge and Valley, but oaks, tuliptrees, and hickories still dominate. American chestnut was formerly important in these forests, too. Old chestnut stumps and rotting logs on the forest floor can still be seen, even though most big chestnuts were killed by the blight more than fifty years ago, a testament to the rot-resistance of this tree, whose wood was highly valued. Piedmont forests are fairly uniform, the only major exception being the floodplain forests, which are extremely rich and diverse. The spring wildflower displays along Piedmont floodplains are one of the continent's greatest floral shows. Floodplains along the Potomac River near Washington, D.C., Bull Run near Manassas, Virginia, and Gunpowder Falls near Baltimore are sites of superb spring wildflower displays.

The forest character changes quickly east of the Fall Line. Coastal Plain forests that develop on this region's rich soils feature white oak, sweetgum, American beech, and other hardwoods in most areas, often with a lot of American holly in the understory. Environmental disturbance on the Coastal Plain gives rise to nearly uniform groves of loblolly pine—tall, long-needled conifers which, in the absence of cyclical disturbance, are eventually replaced by the hardwood forest just described.

Wetlands cover more of the Coastal Plain than any of the area's other provinces. Rivers in the southern Coastal Plain are often slow-moving and broad, periodically flooding wide areas adjacent to the main channel. Swamp forest, dominated by baldcypress, red maple, and a variety of other flood-resistant trees, occurs in these areas. Brackish and salt marshes are frequent along the tidal portions of our rivers, the shores of the Chesapeake, and along the Atlantic coast and its bays, with freshwater tidal marshes found near the head of tide (that is, the uppermost place in a river where tidal flow occurs) on most Coastal Plain rivers. Fresh, brackish, and salt marshes are all dominated by low-growing grasses, rushes, and sedges. Saltmarsh cordgrass, saltmeadow hay, saltgrass, and black needlerush abound in mid-Atlantic salt marshes. All are also found in brackish wetlands, along with olney three-square, narrow-leaved cattail, saltmarsh bulrush, and big cordgrass. Freshwater marshes vary from place to place, but generally they are dominated by some mixture of the following plants: great bulrush, soft rush, umbrella sedge, tussock sedge, American three-square, broad-leaved cattail, narrow-leaved cattail, wildrice, Walter's millet, water dock, buttonbush, arrow-leaved tearthumb, and marsh hibiscus. Calm ponds of fresh water feature emergent plants such as spatterdock, arrow arum, pickerelweed, and broad-leaved arrowhead. Common reed, often referred to by its genus name, *Phragmites*, is an aggressive plant that can invade any type of wetland in our region.

Barrier islands along the Atlantic coast and capes that jut out into the sea are, essentially, big mobile sandpiles. Several distinct plant communities occur on these sands, especially in dune areas where the sand is always moving. Beach grass, sea rocket, beach heather, and only a few other low plants are able to tolerate the moving sands and salt spray of the primary dunes, which are those closest to the sea. Farther back from the ocean are the sands of the secondary dunes, where plant diversity is a bit higher. Plants of the secondary dunes include wax myrtle, bayberry, seaside goldenrod, and dog fennel. A woodland community called maritime forest grows on the most stable sands of barrier islands and capes. The maritime forest features loblolly pine, sweetgum, post oak, scarlet oak, red maple, and black cherry trees, with American holly, greenbrier, and wax myrtle in the understory.

PLANTS

Plants are far more abundant than animals in most ecosystems. It's a matter of simple energetics: a limited amount of the Sun's energy reaches Earth, and only green plants can use this energy directly. To get their energy, animals must feed on plants, or feed on animals that have fed on plants. Only a small percentage of the energy that plants gather from the sun is available to animals, however, as most is used in the plants' living processes. Thus, as my ecology professor taught me, the world is filled more with plants than with animals. I'm glad for this. Many plants are elegant in their beauty and their adaptation to the environment. We're fortunate in the mid-Atlantic to have a rich flora, featuring more than a hundred kinds of trees, more than a thousand different wildflowers, and an assortment of other plants.

Wildflowers

Every mid-Atlantic season has wildflowers, for skunk cabbage's burgundy curve-topped cones pop up from wet soil in rich woods during January, and in December the delicate yellow blossoms of witch hazel still cling to this shrub's leafless twigs. One of our best wildflower shows can be observed during April and May on the floor of rich floodplain forests. Virginia bluebell, spring beauty, cutleaf toothwort, jack-in-the-pulpit, mayapple, wild ginger, bloodroot, Virginia waterleaf, trout-lily, and rue anemone are among the conspicuous wildflowers of this habitat. These early spring wildflowers exploit an ecological niche: the soil and air are warming, day length is long, moisture is usually plentiful, and abundant sunlight passes through the leafless canopy to the forest floor.

Jack-in-the-pulpit

Cold temperatures linger longer in the mountains, so here the greatest numbers of woodland wildflowers bloom during May and June. Some of the same flowers that decorate floodplain forests also occur in the mountains, with additions including large-flowered trillium, Canada mayflower, bunchberry, starflower, foamflower, goldthread, fringed polygala, rose twisted-stalk, and yellow ladyslipper. June through August bring many flowers into bloom in open, sunny areas. Wildflowers of summer meadows include butterfly weed, evening primrose, black-eyed susan, Queen Anne's lace, wild lettuce, New York ironweed, and rough-fruited cinquefoil. Summer triggers the blooming of a few forest plants, including the wild orchids: rattlesnake plantain, large twayblade, round-leaved orchid, spotted coralroot, and cranefly orchid.

Many wildflowers grow in freshwater marshes and calm, shallow ponds. Visit these habitats in summer and you can find the showy flowers of pickerelweed, spatterdock, crimson-eyed rose mallow, swamp milkweed, cardinal flower, lizard's-tail, and white turtlehead. Bogs support several uncommon wildflowers that grow from their peaty soils, including sheep laurel, rose pogonia, pitcher plant, cranberry, meadow beauty, round-leaved sundew, and red milkweed. Rocky cliffs provide habitat for yet another set of wildflowers, including fire pink, columbine, moss phlox, stonecrop, and alumroot.

Trees and Shrubs

Tree lovers are blessed with a great variety of trees to study in the natural areas surrounding Washington, D.C. and Baltimore. The mid-Atlantic forests range from red spruce and balsam fir groves in the West Virginia Highlands, which resemble the boreal forest of Canada, to the baldcypress swamps and surrounding dry piney woods of the Eastern Shore, a look not unlike that of coastal Georgia. Between these regional extremes are several distinct forest types with many different trees to learn and enjoy.

Deciduous forest dominates the region, and in many places the trees hold their leaves for almost exactly half the year, from late April to late October. I've always delighted at this symmetry. For six months we see our forest trees as bare, skeletal outlines, whereas for the other six the entire forested world is a riot of rich greenery. The transitions are punctuated with color, the bold tones of fall's foliage and the subtle pastels of spring's emerging buds and flowers.

Widespread deciduous trees of mature mid-Atlantic forests include black walnut, American beech, basswood, black gum, mockernut hickory, pignut hickory, black birch, white oak, chestnut oak, pin oak, and red oak. Trees better suited to the sunnier conditions of forests in the early stages of succession include red maple, black locust, black cherry, persimmon, serviceberry, hackberry, and sassafras. Virginia pine, loblolly pine, white pine, pitch pine, and Virginia juniper are conifers that also thrive as part of a young forest. Some mid-Atlantic trees are most common in the mountains, including bigtooth aspen, shagbark hickory, yellow birch, cucumber tree, and sugar maple. Others, usually found in our lowlands, include sweet bay, American holly, southern red oak, willow oak, and sweetgum. Trees that most often grow from moist soil near rivers or wetlands include black willow, river birch, sycamore, pawpaw, silver maple, baldcypress, white ash, and green ash. Rocky or sandy dry soils are the places where blackjack oak, post oak, bear oak, chinquapin, and table mountain pine are usually found.

Most of our forests include a nice mix of canopy trees, understory trees, shrubs, and vines. Most of the canopy trees are wind-pollinated, with inconspicuous flowers; tuliptrees and magnolias are exceptions. Showy flowers are designed to attract insects or other pollinating animals. Many smaller woody plants have attractive flowers because wind pollination is less reliable under the shelter of the canopy. During April two of our showiest small trees, flowering dogwood and redbud, bloom in the forest understory. Several shrubs of the heath family are among the best-loved flowers of spring: the wild azaleas, rhododendrons, and mountain laurel, all of which bloom in May through most of the region, in June in the highlands. Common woody vines include trumpet creeper, wild grape, poison ivy, and Virginia creeper.

The intricate pipevine is found mostly in the mountains; it is common in the southern part of Virginia and West Virginia. Yellow jessamine, crossvine, and climbing hydrangea are found in southeastern Virginia. All three are common in the Great Dismal Swamp.

Ferns, Other Plants, and Fungi

The diversity found in wildflowers and woody plants is reflected in the more primitive plants such as ferns, mosses, lichens, and horsetails. Some are evergreen whereas others wither in winter. Most ferns prefer moist areas, as do mosses and most of the less frequently noticed nonvascular plants—quillworts, liverworts, and hornworts. Many ferns are conspicuous in our forests.

Christmas fern, widespread throughout mid-Atlantic forests, derives its name from its evergreen form of growth: Because its leaves are green in winter, it was historically used for Christmas decorations. Two other common evergreen ferns are the intermediate wood fern and the marginal shield fern. Deciduous ferns that are widespread as forest understory plants include broad beech fern, New York fern, sensitive fern, ebony spleenwort, and lady fern. The big and robust cinnamon fern occurs in many areas where the soil is damp or wet, sometimes intermixed with one of its close relatives, royal fern or interrupted fern. Two ferns specialize in open, sunny habitats: bracken fern and hay-scented fern, which sometimes forms a dense carpet in open mountain meadows. Rock polypody grows on bare rocks throughout the region. A group of small ferns, including walking fern, American wallrue, and purple cliffbrake, only grow on limestone cliffs. Others, such as Allegheny cliff fern, hairy lip fern, and mountain spleenwort, are more often found growing from dry shale or sandstone cliffs and rockpiles.

Fungi used to be considered a specialized group of nonflowering plants, though now many specialists place them in their own group of organisms, not strictly plant or animal. However you categorize them, mushrooms and other fungi are conspicuous parts of our field and forest ecosystems. They are important decomposers in the soil, breaking down organic material into simple compounds that can then be used by plants. The bulk of any fungus is its mycelium, a mass of threadlike strands that run through the soil or the rotting wood on which the fungus exists. Still, it's the fruiting body of a fungus that most people notice in the field; most of these fruiting bodies are called mushrooms.

Some mushrooms can be found at any season, and a few kinds of woody shelf fungus can last for many years, but late summer and fall tend to be the seasons of greatest mushroom abundance and diversity. Look for meadow mushrooms, puffballs, and inky caps in open grassy fields. Many different mushrooms can be found in our forests when the soil is very moist after long rainy spells. Some of the most distinctive woodland mushrooms are the pale red russulas, the big spongy-capped boletes, the rich golden chanterelle, which is edible, and the quite similar jack-o-lantern, which is poisonous. The jack-o-lantern gets its name from the soft glow that is emitted from its gills, visible only on very dark nights. Morels, whose convoluted caps can be described as brainlike in appearance, are springtime mushrooms that are highly prized for their rich, earthy flavor.

Fungi take many forms besides that of the familiar mushroom. Stemless shelf fungi grow straight out from the trunk of a standing dead tree or a fallen log. Two of our most conspicuous shelf fungi are the finely banded turkey-tails and the vivid orange and yellow sulphur shelf. Some odd-looking fungi have names that describe their unusual appearances, such as dead-man's fingers, scarlet cup, orange peel, earthstar, and witches' butter. One of the strangest-looking fungi is the stage of the cedar apple rust that grows from the branches of juniper trees. It looks like a small brown golf ball with slimy orange worms crawling from it. This same fungus, at another stage, forms a pale orange-brown bloom on the leaves of apple trees; it is considered an agricultural pest.

Some fungi are tasty and highly prized, but a few are highly toxic and can cause death if eaten. Fungi present some tricky identification challenges. I encourage everyone to enjoy the beauty of mushrooms in our fields and forests, to appreciate their importance in our ecosystems, and to learn about their identification. Don't eat any wild mushroom, however, unless you've learned its identity from a reliable expert.

Lichens are unique organisms in which a fungus and an alga live in union. The fungus forms the structure, whereas the single-celled alga produces the energy. Algae, like most other plants, convert solar energy to chemical energy through the process of photosynthesis. No fungi contain chlorophyll, the chemical required for photosynthesis. The algal component of a lichen produces enough energy to support itself and its fungal partner.

There are three types of lichens. Crustose lichens grow like a crust on rocks or logs, their cells intermingling with those of the substrate on which they grow. It's impossible to remove an intact crustose lichen from a rock or log unless part of the substrate is peeled away, too. Foliose lichens take a leafy form. Although some foliose lichens are attached snugly to rocks or logs and resemble crustose lichens, foliose lichens peel readily away from their substrate. Fruticose lichens have a branching structure; some have short, leafy branches whereas others have long, stringy threads.

WILDLIFE

The mid-Atlantic region is home to a great variety of wildlife, thanks to its diverse environments and plant communities. Although most people generally love wildlife—witness the popularity of television nature programs and zoos—not everyone likes every wild animal. People often apply the term wildlife only to those wild animals they want to see; they may call other animals pests. I try not to be judgmental. To me, a discussion of wildlife encompasses all the living animals that are part of our environment, except for those which are domesticated. Bobcat, mockingbird, bullfrog, rattlesnake, rainbow trout, earthworm, dragonfly, and mosquito all qualify as wildlife. The next sections provide an overview of the wildlife with whom we share the mid-Atlantic region. I begin with discussions of the five groups of vertebrates, those animals with backbones—birds, mammals, reptiles, amphibians, and fish—and then provide an overview of invertebrates, those animals without backbones.

Birds

Birds are surely one of the most charismatic groups of wildlife, with more people taking an interest in birds and birding than in any other natural history pursuit. Not only are birds often visually beautiful, but many also make marvelous sounds and engage in fascinating behaviors. Birds are also abundant, diverse, and fairly easy to observe and identify. Fledgling birders can turn for help and information to excellent resources including books, field guides, recordings, and computer programs as well as clubs and nature centers that offer field trips with knowledgeable leaders.

Although many people observe birds in casual ways, learning only the most common and conspicuous species, many others discover that birding soon becomes an overriding passion. Many ardent birders keep a life list—

a record of all the bird species they have seen in their lives. A bird seen for the first time and added to one's life list is called a life bird. Area birders eager to build a large life list—say 700 species for the United States and Canada—need to see almost all of the area's nesting birds plus some strays that don't normally occur in this area.

Catbird in tuliptree

Because many species are strong fliers, birds are capable of moving to parts of the world far away from their normal ranges. Every once in a while, birds that live in Europe or Asia somehow find their way to North America (and vice versa). When a bird is found outside its normal range it is called a rarity. The quest for such rarities is a popular activity among enthusiastic birders everywhere. The mid-Atlantic has a fairly good history of rarities, especially along the coast. *The Voice of the Naturalist* is a weekly recording prepared by Audubon Naturalist Society volunteers that highlights unusual bird sightings in the mid-Atlantic. Call 301-652-1088 to hear the latest birding news. When an especially unusual bird is found and reported on the tape, hundreds of excited birders may converge on its location. Comments on birds in this book will primarily focus on our regularly occurring species, but I do note some rarities that have been seen.

A variety of birds may be observed any day of the year in every part of the mid-Atlantic. Knowledgeable birders can find at least 50 species a day if they visit several habitats, including at least one with water. Birders who make a serious effort can usually find more than 300 species in our region over the course of a year. Much discussion about seasonal highlights and regional hot spots for birding are included in subsequent chapters of this book, but a brief summary is included here.

Winter birding is best near water, where ducks, geese, swans, loons, and other waterbirds are plentiful throughout the coldest months. Seed-eating songbirds are easily observed during winter at bird feeders. Occasionally birds whose ranges are normally farther north come to the mid-Atlantic in winter, including evening grosbeaks, red crossbills, common redpolls, northern shrikes, and snowy owls.

Spring migration is spread over March, April, and May, but the mass movement of songbirds—a movement that is one of the region's greatest birding phenomena—runs from April's last week through about May 20. Each year during this period a few small teams of experienced birders spend an entire day afield in search of the greatest variety of birds. In birding lingo this is called a big day. The best big day totals from the mid-Atlantic region exceed 200 species, and every day brings a new set of birds into the mid-Atlantic. Migrant songbirds include many warblers, vireos, flycatchers, tanagers, orioles, and sparrows.

Summer features nesting songbirds and the year's best opportunities to observe herons, egrets, and terns. The fall migration of shorebirds begins in July, whereas the southbound movement of songbirds and raptors begins in August and peaks in September and October. November brings the arrival of great numbers of ducks, geese, swans, grebes, loons, and cormorants into our region. Most of these birds will spend the winter here, heading north again in March and April.

Mammals

We humans tend to identify most closely with mammals because we are mammals ourselves—warm-blooded vertebrates whose females nurture newborn young with milk produced in mammary glands. Mammals have hair-covered skin, as many as four kinds of specialized teeth, and almost all bear live young. Most have well-developed limbs and a highly developed sense of smell. Most people thrill to the sight of a rarely seen wild mammal, such as a black bear, bobcat, or flying squirrel. We become more nonchalant about those more easily observed mammals, such as white-tailed deer and eastern gray squirrel, yet these animals give us great opportunities to study the ways mammals feed, court, care for young, and generally interact with their environment.

Most of our region's wild mammals are secretive and nocturnal (more active at night than during the day), making it difficult to make a search for mammals your primary focus on a nature excursion. Most naturalists try to stay constantly alert for the presence of mammals, primarily by relying on their signs, such as tracks, droppings, and other evidence of their travels. To study small mammals in the field, scientists generally set out large numbers of live traps in suitable habitat. Winter is an excellent time to search for mammal signs, especially when there is a light dusting of snow. Of course we've got a few species that won't be found in winter because they either hibernate or migrate; these include woodchucks, chipmunks, and many species of bats.

The ecological definition of carnivore is an animal that feeds on living animals. In mammalogy carnivore also refers to a specific group of related mammals, the order Carnivora. Some of these mammals are ecological carnivores, whereas others are omnivores, feeding on a variety of foods, including living animals. Carnivores include bears, cats, dogs, raccoons, weasels, and seals. Many are uncommon. For example, black bear and bobcat only survive in remote areas with little human disturbance. Mountain lions may still exist in our wildest mountain areas, though no recent sightings have been confirmed. Coyotes have only recently moved into the mid-Atlantic; they are still quite uncommon. Based on their success in other regions, coyote populations are expected to increase.

Red fox and gray fox are fairly common and widespread throughout our region, though foxes usually behave secretively and are difficult to see. Raccoons are common and often grow accustomed to humans, making them more easily observed than other carnivores. Members of the weasel family include the widespread striped skunk, spotted skunk and least weasel, which live in the mountains, and the secretive long-tailed weasel, which lives throughout the eastern United States yet is rarely seen. Populations of river otter and mink are slowly rebounding from heavy trapping of the past. Either may be seen near water in our region, though both are still uncommon.

Large hoofed mammals are important members of many wildlife communities. In our region only one wild species, the white-tailed deer, fills this niche, though historically both elk and bison lived in the mid-Atlantic. These last two species were removed from our region because of habitat modification and hunting by those settlers who cleared lands and built homes and farms during the eighteenth and nineteenth centuries. Deer, which were also almost eliminated, are now abundant and easy to see throughout the region, though as recently as the 1960s they were still uncommon. In some places deer are now considered pests because they damage crops and gardens, overbrowse shrubs and young trees in forests, and create traffic hazards on roads by leaping suddenly in front of oncoming vehicles.

Rodents are a large group of mammals that includes mice, rats, voles, and beavers. In much of the mid-Atlantic region our most readily observed mammal is a rodent—the eastern gray squirrel. We tend to take these agile animals for granted, yet much can be learned about behavior and adaptation by watching squirrels. Fox squirrels are larger than gray squirrels and far less common in this area. One subspecies is endangered—the Delmarva fox squirrel, which is only found east of the Chesapeake Bay in Maryland, Delaware, and Virginia. Red squirrels are smaller than gray squirrels and far less common, though good numbers may be found in the conifer forests of our highest mountains. Southern flying squirrels are common and widespread, but not often seen because of their strictly nocturnal habits. Eastern chipmunks, abundant forest dwellers, are the smallest member of the squirrel family in eastern North America.

Three species of rodents that are not native to our region—black rat, Norway rat, and house mouse—are significant pests. We dislike them all when they move into our houses and damage property. Unfortunately their bad reputation has led to a widespread dislike of our native mice and rats, which rarely interact with people. Many small rodents live in mid-Atlantic fields, forests, and wetlands, including harvest mice, deer mice, white-footed mice, meadow voles, southern red-backed voles, eastern woodrats, and marsh rice rats. Many are important prey for carnivorous birds, reptiles, and mammals.

The beaver is a rodent with a better reputation, but as this mammal's population increases, its reputation may be declining. Like many other mammals, beavers were almost eliminated from the mid-Atlantic during the nineteenth century because of habitat destruction and trapping. During the first half of the twentieth century, beavers were limited to a few areas in the high mountains. When beaver trapping was finally restricted, their populations began to recover. Wildlife managers reintroduced beavers into waterways throughout the mid-Atlantic, and by 1980 beaver populations had reestablished in many areas.

Beavers change the landscape more than any other wild animal—family groups of beavers will build dams made of small trees and mud across streams, creating small ponds. They gnaw through the base of trees, cutting them down for food or building material. The ponds they create provide these semiaquatic animals with shelter, transit corridors, and food. Aquatic vegetation of the pond and surrounding wetlands is an important summer food source for beavers, which feed primarily on tree bark in winter. Although beaver ponds are considered scenic treasures in parks and other natural areas, they are less appreciated when they flood farm fields, suburban backyards, or roads. As more beavers establish territories near people, sometimes cutting down prized trees for food or dam-building material, they may be considered pests.

Two other rodent species, muskrat and nutria, also live semiaquatic lives, though neither builds dams. Muskrats are much smaller than beavers, relatively common in wetlands throughout the mid-Atlantic, and are abundant in the wetlands that border the Chesapeake Bay. An average adult muskrat weighs about 2 pounds, whereas a big beaver could weigh 60 pounds. Nutria are large rodents—as heavy as 20 pounds—native to South America. These animals have been successfully introduced into wetlands along the Chesapeake and in other parts of the United States. Nutria have become problematic in many areas, as is all too often the case with introduced animals. They tear apart marsh vegetation and diminish the ability of a wetland to provide food and shelter for other animals.

Many other mammals are found in the mid-Atlantic region. Virginia opossum, eastern cottontail, and woodchuck are common. Nine species of shrews and three species of moles also occur here. None are easy to observe without trapping. Twelve species of bats live in our region; two, Townsend's big-eared bat and the Indiana bat, are federally listed endangered species. Although it's easy to see bats circling overhead at twilight, it's nearly impossible to determine species by such fleeting glimpses. Bats roost in sheltered places such as caves, hollow trees, or buildings. If you find a roosting bat it will most likely be a little brown myotis, eastern pipistrelle, red bat, evening bat, or a big brown bat, for these are our most common species.

Several marine mammals may be observed in the Atlantic Ocean and occasionally in the Chesapeake and other coastal bays. Only the Atlantic bottlenose dolphin is regularly seen near land, though twenty-nine species of whales and dolphins have been recorded in the waters of the Atlantic between southern New Jersey and Virginia. The harbor, harp, and hooded seals have been seen in our region, though none are regularly observed here.

Reptiles

Three major groups of reptiles—the snakes, turtles, and lizards—are well-represented in the mid-Atlantic fauna. All are cold-blooded animals, so in winter they become dormant and hide in sheltered places. Summer's warmth brings our region's reptiles into a high level of activity, though sunny warm days in spring and fall can be superior for reptile watching, as many then bask in open places to gather up the sun's warmth. These are the days when a search for reptiles is likely to be most successful, though it's not unusual to see two or three species on any given field day between April and October.

SNAKES. Snakes fascinate many people; some people are intrigued, others are fearful. Snakes have suffered terribly from people who don't understand them. Because a few species are venomous, some people routinely try to kill any snake they might find. This is a shame because snakes are important predators of small animals, and most are harmless. Many are beautifully colored, and all are fascinating to watch. Their ability to move quickly by sliding their overlapping belly scales is amazing to observe. Many nonpoisonous snakes may be carefully picked up and handled, though some can give a painful bite. Most naturalists prefer to leave snakes untouched and simply watch their activities.

Snakes employ many different strategies to catch prey. Some hide quietly under vegetative cover, hoping to ambush a passing animal. Others climb through the branches of trees and shrubs, probing the air regularly with their forked tongues, which are sensitive organs of smell. These snakes may be searching for young birds or eggs, for young squirrels, bats, or chipmunks, or for large insects. Snakes are often seen around homes and cabins in rural settings, where they search for mice and rats. Many snakes are fine swimmers and hunt for frogs, toads, fish, and aquatic invertebrates. Most mid-Atlantic snakes mate in April or May. Some lay eggs whereas others bear live young. Newborn and newly hatched snakes are most often seen during August and September.

Even though three species of poisonous snakes occur in the mid-Atlantic, their bites are quite unusual and rarely fatal. Most bites occur after a person tries to pick up and hold a poisonous snake, a practice I definitely don't recommend. Whenever you search for salamanders or insects under rocks and logs, however, stay alert to the possibility of finding a poisonous snake because these are among their favorite resting places. The copperhead, our most common poisonous snake, is widespread but infrequently seen because most are rather small and all are timid and secretive. Timber rattlesnakes are fairly common in the mountains. They are found most frequently around rockpiles, though on sunny days they may be found basking in any open area. Many years ago I nearly stepped on a big, beautiful rattlesnake stretched across the Appalachian Trail in Shenandoah National Park. I have never jumped backward as quickly as when I heard the distinctive buzz of that rattler's tail, a moment I still vividly remember. A different subspecies of the timber rattlesnake, usually called the canebrake rattlesnake, occurs in southeastern swamps, reaching into Virginia at the Great Dismal Swamp. The cottonmouth (also called the water moccasin) reaches the northern limit of its range in southeastern Virginia. These heavy-bodied snakes will sometimes open and close their mouths rapidly, flashing the bright white lining inside as a warning. The first time I saw this from the corner of my eye I thought I was seeing a white butterfly opening and closing its wings! Fortunately, I corrected my identification before I moved in for a closer look.

Nonpoisonous snakes are seen far more often than poisonous ones. The black rat snake is frequently seen around human dwellings in suburban and rural settings. Eastern garter snakes are common in woodlands throughout the region. The northern water snake is abundant wherever there is quiet water. Some of our snakes are beautifully colored yet rarely seen: The rough green snake, scarlet snake, corn snake, and scarlet kingsnake all sport bright colors. Our smallest snakes are most often found by looking under rocks, logs, old boards, or other debris. Ring-necked snakes are never longer than 20 inches, and the biggest brown snakes are just 18 inches. Our three smallest snakes, the smooth earth snake, redbelly snake, and worm snake, are usually less than one foot in length.

TURTLES. Many mid-Atlantic turtles are easy to see around areas of quiet water. The snapping turtle is our most imposing species. A big snapper, with its spiked tail and fearsome look, would seem to have come from the dinosaur era. At the gentle side of the spectrum is the box turtle, whose elegantly hinged shell provides a nearly perfect shelter from predation. The box turtle is the only common turtle of our region that regularly travels far from water. Through most of the mid-Atlantic, the painted turtle is the species most commonly seen basking on floating logs and other debris in quiet ponds and marshes. Red-bellied sliders are often found with painted turtles in the Piedmont and Coastal Plain. Harder to find are the eastern mud turtle, spotted turtle, river cooter, and stinkpot. These turtles spend most of their time under the water and hidden from view, rarely basking in the open. Populations of diamondback terrapins, which live in salt marshes, are slowly rebounding from days when these turtles were commercially harvested for turtle soup. Loggerhead sea turtles are seen on rare occasions in the mid-Atlantic region, either in the open Atlantic Ocean or in one of the coastal bays. Like all sea turtles around the globe, the loggerhead population has plummeted because of habitat degradation and overharvesting of adults and eggs.

LIZARDS. Lizards are less common here than in arid regions of the American West, but a few species occur, most often in dry forests. The five-lined skink is most frequently seen; females are dark with five yellow lines running along the body from the snout to midtail. Immature five-lined skinks look much like females, except their tails are bright blue whereas the tails of adult females are black and partially striped. Adult males are coppery-red and unstriped. Five-lined skinks are generally between 4 and 8 inches long. Broad-headed skinks look very much like five-lined skinks, though this species can be much longer, up to 13 inches. Female and young broad-headed skinks have seven longitudinal stripes. Unfortunately, whenever I find a skink it seems to be dashing away too quickly to count the stripes!

The eastern fence lizard is the only other lizard I have seen in the mid-Atlantic region. These plump, gray-brown lizards are marked with uneven white and black stripes, and inhabit open, dry forests and rockpiles. Other lizards that occur in our region are the coal skink, ground skink, and the six-lined racerunner. Two other species may be found in southeastern Virginia: the southeastern five-lined skink and the snakelike slender glass lizard, whose tail "shatters" when a predator tries to catch it.

Amphibians

FROGS. Nothing signals the beginning of spring in the mid-Atlantic as emphatically as the raucous chorus of spring peepers, which generally begin calling in mid-March. These tiny frogs have big voices, especially when they congregate in quiet freshwater ponds and pools and sing together through the night. They're not the first frog to become active in spring, however. The wood frog, a gray-green frog with a black mask, generally breeds in February! Throughout spring and into summer, frog song is a conspicuous part of our natural world. Most frogs sing during the night, though some song can be heard at any hour. More species start singing as spring progresses, with April and May being peak months. During the peak of their breeding cycle most frog species are easy to find at night in a flashlight beam. At other times, day or night, finding frogs takes some vigilance and some luck, for they blend well into their surroundings and are usually quite wary.

Frogs, like other cold-blooded animals, cannot stay active during the coldest months of the year. They spend the winter in a dormant stage in moist, sheltered places such as the bottom of ponds, an underground burrow, or beneath the forest leaf litter. For much of the year they are secretive, their camouflage generally quite effective. During spring frogs mate, and sound is an important part of the ritual, making springtime the ideal season for frog observation. Compared with birds and flowers, our region's diversity of frogs is low. Most species are quite abundant, however, so it's not too hard to learn their sounds and identifying characteristics.

Wood frogs are usually the first frogs to be found each year. A spell of mild winter weather—especially one with rain—during February or March can bring wood frogs out of their winter dormancy and into a brief but torrid mating period. Wood frogs winter throughout the forest, but upon emergence they head directly to bodies of standing water, either permanent ponds or temporary vernal pools. Here the males will give their sharp calls, which sound a bit like a cluck or a quack, and aggressively try to mate with anything resembling a female wood frog.

A male frog mates by swimming up to a female from behind, grasping her firmly with his front legs and hanging on. It looks like an aquatic game of piggyback. Fertilization is external; when the female chooses to secrete her eggs, the male fertilizes them in the water. The mating fervor is so great that sometimes male wood frogs will also aggressively clamp onto other males and even bits of floating debris that don't even vaguely resemble a frog. I've seen up to five male wood frogs chained together to a female. Perhaps those farther back in the chain also succeed in fertilizing some of the eggs. The entire population of wood frogs in a given area will complete its mating cycle within a few days. If you visit woodland ponds at the right time, you're certain to see and hear wood frogs. If you miss it, you'll have to wait until the following year.

Spring peepers become conspicuous members of our wildlife communities each March when their breeding season begins. These little treefrogs are only about an inch long, yet they have extraordinarily powerful voices. Their calls are shrill, upslurred whistles. On early March nights a few peepers can sometimes be heard near quiet waters. As the month progresses, waters warm and day length increases, bringing more peepers out of dormancy and into the courtship ponds. Their mating concentrations peak in late March or early April, and an evening visit to a pond, pool, or marsh at this season is a noisy event. With dozens, or even hundreds, of spring peepers calling simultaneously, each trying to call more loudly than the next, the resulting choral performance is incredibly loud. More than once I've driven past a wetland at this season, windows rolled up in my car and music playing over the radio, and I have still distinctly heard the chorus of spring peepers.

Singing frogs are generally quite aware of their surroundings, and they will stop singing with the approach of large mammals like ourselves. At the frenzied peak of spring peeper courtship, however, they can't stay quiet for long. Walk down to a peeper pond on a late March evening, stand quietly for a minute or two, and the songs will no doubt begin again. With a good eye and a decent flashlight you should be able to find some singing males, their throat pouches inflated like bubbles of gum. You may find them squatting at the water's edge, resting on floating vegetation, or clinging to vegetation several feet above the ground. If you're sensitive to loud noises, wear earplugs.

As spring progresses, spring peepers continue to call, though less stridently and in lesser numbers. Other frogs whose mating seasons begin in March reach their peaks in April and May. From late March through early June you may hear the low-pitched, buzzy noise of a pickerel frog, an attractive frog that has rather square, olive-brown spots set against a dusky gray background. You might also hear the rhythmic grunting of a bullfrog, our largest frog, which sometimes reaches 8 inches in length. Bullfrogs are variably gray-brown above, with bold, bulging eyes and a dorsilateral fold, the prominent ridge of skin that extends back from the eye to the top of the ear and then curves sharply downward.

Another common sound of wetlands in spring is the sharp twang of a green frog, which looks a bit like a small bullfrog. To be sure of your identification, check this frog's dorsilateral fold, which extends straight back from the eye to the base of its hind legs. You may hear the long, steady, high-pitched trills of American toads near wetlands, but these toads also call from dry patches of forest. Toads are easy to distinguish from frogs by their plump appearance and the raised, warty lumps found on their skin. American toads continue to call at night throughout the summer. The American toad generally has dark splotches on the back that correspond with the warts. Its cousin, the Fowler's toad, is common on the Coastal Plain and generally has three or more small warts in each dark blotch.

You may hear the gray treefrog around a pond at night, though you're just about as likely to hear these 2-inch-long tree climbers during the day. Their rather harsh trill bears some resemblance to the call of a red-bellied woodpecker. I hear them calling a few times a day during the spring bird migration in late April and early May. Gray treefrogs are a lovely mottled gray and look much like a lichen-covered lump on a tree branch.

These are the more common frogs of the immediate Washington-Baltimore area. Like many other nature study pursuits, frog watching (and listening) becomes addictive. Once you've learned the residents of your local pond, you might want to venture into new habitats in search of other species. Freshwater marshes in the Coastal Plain are great spots for springtime frog searches. I've also had great frog-watching fun along the swamps of the Pocomoke River, the marshes on the Elliott Island Road, and at the Great Dismal Swamp, the only place where I've heard carpenter frogs—named for their call, which sounds like someone hammering nails into wood.

Sadly, frog populations worldwide seem to be declining. Habitat loss and degradation of remaining habitat quality are surely major reasons for their decline. There is some evidence that their eggs are sensitive to ultraviolet radiation, and so the damage we've caused to the atmosphere's ozone layer may be another reason for this decline. Other causes may include climate changes, nonnative predators, pesticides, and acid precipitation. Fortunately there are still good numbers of frogs in many of our nearby wetlands.

SALAMANDERS. Eastern North America has the world's greatest diversity of salamanders, and the mid-Atlantic region is home to many of these slender, quiet amphibians. Like frogs and reptiles, salamanders are cold-blooded, so you're not likely to find them in winter. Some breed very early in the spring. If you visit a pond full of spring peepers in March, scan the bottom of the water and the surrounding moist ground for the big, bold spotted salamander, which breeds at about the same time and place. Salamanders are active from about March through November in most mid-Atlantic areas.

Salamanders look a bit like lizards, but they differ from those reptiles in many ways. Lizards have claws on their toes, external ear openings, and scale-covered skin; salamanders lack all of these. Lizards breathe air for their entire lives; salamanders, like all amphibians, spend part of their lives obtaining oxygen from water through gills. Salamanders are primarily nocturnal, and most need to keep their skin moist at all times, so the ideal time to look for them is on a rainy night. During the day many salamanders may be found resting under rocks or logs in moist forests or along streams.

Our most common species regionwide is the red-backed salamander, which comes in two color forms, one dark with a red back and the other all dark; this color form is sometimes called the *lead-backed*. Both forms have blotchy gray and black bellies, though they're generally not fond of showing off their underparts! Spotted salamanders have large black bodies decorated with big, round, yellow spots on the back. Each has a unique spot pattern. Spotted salamanders are common in the mid-Atlantic, though for much of the year they stay well underground and are hard to find. This changes abruptly in early spring, at about the same time that spring peepers are beginning to sing. Then spotted salamanders emerge and travel to quiet ponds and vernal pools, where they sometimes congregate in groups of 100 or more. All swim vigorously around these little ponds in preparation for mating. Although males and females gather together in these aggregations, they don't actually contact one another to mate. Males deposit spermatophores—little packets of jellylike material that include their sperm—all around the pond. Females select a spermatophore and settle onto it, resulting in fertilization of her eggs. She lays about a hundred eggs in the same pond within a few days. A few days later all the breeding spotted salamanders leave the pond and return to the surrounding forest. The eggs hatch a month or two later in the breeding ponds, where the gilled larvae will feed on aquatic insects and other small animals. Many larvae are eaten by larger animals, but those that survive transform to the adult stage in late summer or early fall.

Some kinds of salamander are almost always found near ponds or streams. Check under rocks along a streambed for northern two-lined, northern dusky, and spring salamanders. Under nearby logs and stones you may find slimy, marbled, or red salamanders. Visit the mountains to search similar habitat for long-tailed, Wehrle's, and Jefferson salamanders. Head to the quiet waters and swamp forests of the Coastal Plain to seek three-lined, mud, and tiger salamanders.

Several species of salamanders are found in extremely limited sections of the Appalachian Mountains, and some of these are listed as threatened or endangered species. The Cheat Mountain salamander is found only in the highlands of West Virginia's Allegheny Plateau. This small salamander is black with many tiny bronze-colored specks on its back. The Shenandoah salamander, which looks a lot like the red-backed, is known only from rocky scree areas on the slopes of the highest mountains of Shenandoah National Park. The Peaks of Otter salamander is found only at the Peaks of Otter, a mountainous area near Roanoke, Virginia. This is another small salamander, usually just 3–5 inches long. Its back is a pale mustard or copper color, and its sides and belly are black with a few white speckles. Although I'm always delighted to find any salamander, to successfully search for one of the rare species is especially gratifying.

Fish

A prominent feature of our region is its tremendously varied set of aquatic habitats, which include cool mountain streams, warm southern swamps, gentle coastal rivers, ponds, lakes, marshes ranging from fresh to salt, estuaries, and the ocean. Each habitat supports a rich and distinctive community of plants and animals, including fish. Many people like to catch fish, and some also like to learn about them and their role in the ecosystem, but there aren't too many fish watchers. This may be changing, however, as there is a growing interest in the general study of aquatic ecosystems, including the small fish known as minnows. Various small netting devices, including dip nets, seines, and minnow traps, enable observers to look at fish and other aquatic creatures. Some fish can be studied by observing clear waters by eye or with binoculars. To look for fish, some people take to the water wearing a diving mask and either a snorkel or a scuba tank. An angling rod and reel are useful tools for getting a look at bigger fish. Pinch the barbs off your hooks if you're interested in just observing fish, and you'll be less likely to injure them.

Small fish are called minnows. When I was young I assumed all minnows were the same, and that their only value was as bait for catching bigger fish. In truth there are hundreds of species of fish in the United States, and thousands around the world that might be called minnows, their maximum length being no more than 2 or 3 inches. Some are widespread, whereas others are so specialized that they can survive in only a single river system. Many of these specialized species are now listed as threatened or endangered. Threats to fish include water pollution, construction of dams and other physical barriers along rivers, destruction of habitat, and competition from introduced species. Game fish are regularly introduced into waters where they don't naturally occur to provide sport for anglers, and the presence of these nonnative species often negatively affects populations of native fish.

Fish diversity in the mid-Atlantic is high. Common minnows of freshwater habitats include black-nosed dace, rosy-sided dace, tessellated darter, shield darter, fantail darter, fathead minnow, mottled sculpin, and creek chub.

In salt and brackish waters, rainwater killifish, banded killifish, fourspine stickleback, mummichog, and sheepshead minnow are among the most common small fish. Many larger fish are eagerly sought by the many anglers of our region. Common sport fish of mid-Atlantic freshwater habitats include largemouth bass, smallmouth bass, brook trout, brown trout, rainbow trout, pumpkinseed sunfish, yellow perch, white perch, and several species of catfish.

Black-nosed dace

Many other large fish may be caught in the salt and brackish waters of the mid-Atlantic, including bluefish, Atlantic menhaden, spotted seatrout, Atlantic croaker, black drum, red drum, black sea bass, flounder, and striped bass. The latter, known locally as rockfish, have long been favored sport fish for many Chesapeake Bay anglers. Populations of striped bass in the bay dropped dramatically in the late 1980s; a fishing ban was put into effect for several years, resulting in a partial recovery.

Some fish spend part of their lives in fresh water and part in salt or brackish waters. Fish that move into fresh waters to breed but spend the rest of their lives in salt water are called anadromous. The best-known anadromous fish of our region is the Atlantic shad. Once an abundant species that provided an important food resource for Native Americans, shad are now absent from many rivers and their populations are depleted in others, largely due to dams that block their movements and water pollution that makes many streams unsuitable for their reproduction.

Invertebrates

Insects and other invertebrates are abundant in general, though many species are uncommon and some are endangered. Many people use the discomfort caused by a few species of insects as reason to disdain this entire group of animals. Of course, I vigorously disagree with this point of view! The economic arguments aside—without insects our agricultural system would collapse and many other elements of our civilization would suffer—insects and other invertebrate animals make for fascinating study. Train yourself to look closely for insects, perhaps armed with a hand lens or magnifier, and you'll find an amazingly rich world. Incredible colors, bizarre adaptations, wondrous behaviors, and elegant interrelationships are all wonders of this miniature world.

Certain groups of insects are more conspicuous than others. Most large insects you find in the field will fall into one of twelve major groups. Four are associated with aquatic habitats: dragonflies, caddisflies, stoneflies, and mayflies. The other eight are beetles, grasshoppers, wasps, flies, true bugs, leafhoppers, lacewings, and moths. Beetles are a diverse group that includes ladybugs, fireflies, weevils, scarab beetles, and long-horned beetles. All beetles have hardened forewings used for protection and not for flight. Grasshoppers, crickets, katydids, walking sticks, roaches, and praying mantids make up a group called Orthopterans, insects noteworthy for their interesting appearance and, for some, their sounds. A late summer evening in our region is almost defined by the chorus of crickets and katydids that provides a universal background drone after dark from late July until the first frost of late October or November.

Wasps, bees, and ants make up another major group of insects. Since many bite or sting, this group is much feared by people, but many species are harmless to people and are important ecologically—pollinating plants, providing food for many birds and other animals, and controlling the populations of other insects. The diverse group of insects called flies are united by a single character—all have just one pair of wings, whereas most insects have two pairs. Many, such as the deer fly, horse fly, black fly, and mosquito, are considered pests because they will bite people, but some of these and many other species are harmless and beneficial.

To an entomologist, *bug* is a term referring to a single order of insects, the Hemiptera, and not to insects in general. This group of insects has forewings that are partially hardened like those of beetles, yet parts of their wings are lacy and used for flight. Widespread members of this group include stink bugs, assassin bugs, milkweed bugs, and boxelder bugs. Hoppers, which include aphids, leafhoppers, and planthoppers, feed on plant juices and most are very small, with cicadas the dramatic exception. Lacewings and their relatives, which include antlions, dobsonflies, fishflies, and hellgrammites, have four lacy wings of roughly equal size as adults. Most are predators that feed on other insects. To trap ants, immature antlions dig conical pits into sandy soil.

BUTTERFLIES AND MOTHS. Of all the insects, butterflies are the best known and most loved. Naturalists have long been fascinated by the beauty and grace of butterflies, but for many years the study of these scale-winged insects was equated with collecting. Although I appreciate the many values of scientific collections, I never relished the prospect of dispatching these lovely animals. I also didn't want the responsibility of caring for a collection, which takes a lot of detail work that my clumsy hands make difficult. So for many years I casually enjoyed the conspicuous butterflies I would see, learning the names of just the few most common species. Then I made a great discovery—butterflies can be studied in other ways. They can be watched while they bask in the sun or feed on wildflowers, and they can be gently netted, handled carefully, and released back into the wild unharmed.

A wonderful variety of butterflies thrives in the diverse habitats of the mid-Atlantic. I've now developed a real butterfly-watching addiction, and I've watched many friends get hooked, too. Butterflies can be found almost everywhere during the warmer months, and almost everyone is charmed by their delicate and intricate beauty. Their remarkable metamorphosis from caterpillar to butterfly has an abiding place in our folklore and traditions. Add a few good nectar sources to any little garden, and a good variety of butterflies are likely to visit.

A few species of butterflies overwinter as adults. The mourning cloak, comma, and question mark may be seen flying through open woods on any warm midwinter day. For the most part, however, the butterfly season runs from April into October. In general, butterflies are short-lived, and those whose life cycle includes just one generation of adults per year will only be seen during a brief portion of the year. A number of such single-brooded species appear in the spring. If June arrives and you haven't seen a falcate orange-tip, henry's elfin, or silvery blue, forget about seeing them for the year and try again the next spring. Other species are multiple-brooded and may be seen throughout the warmer months; these species include the tiger swallowtail, silver-spotted skipper, and pearl crescent.

Most butterflies are restricted to certain types of habitat, and some are closely linked to areas where a specific larval host plant grows. Many young butterflies, that is, caterpillars, will only feed on a single species of plant or a small group of closely related plants. To find some of these butterflies it is necessary to visit areas where the larval host plant occurs. Because of this, butterfly watching can hold quite different rewards in different places and at different times of the year. A little homework can show you when and where to search for many kinds of butterflies. Armed with this knowledge, I delight in going out and trying to find one or more of these highly specialized butterflies. I don't always succeed, which makes the times I do seem especially rewarding.

Moths range from the confusing small brown ones to some of our most incredible insects, such as the luna moth and the agile sphinx moths, whose ability to hover and drink nectar from flowers explains why they are also called hummingbird moths. Immature moths and butterflies are caterpillars, and some are quite conspicuous, such as the wooly bear of autumn and the bizarre hickory horn devil of summer.

AQUATIC INVERTEBRATES. Pick up a rock from the bottom of a clear stream and you may see a young salamander or a crayfish scooting away. Look closely at the rock and you're likely to find one or more insects, probably the immature nymph of an insect that leaves the water for its adult stage. There's a wide assortment of insects that are aquatic as immatures, including dragonflies, damselflies, mayflies, caddisflies, stoneflies, and some beetles. Caddisflies are my favorites; their nymphs use stones or bits of plant material to build little cases in which they live. Those who enjoy fly fishing are advised to learn the insects of any stream they plan to fish, for their success with catching trout depends on using a lure that mimics an insect that occurs in the stream.

Dragonflies and damselflies are often conspicuous in the air above fresh waters during the warmer months. These insects come in many different patterns and colors, including blue, green, orange, red, and black. Field identification of Odonates, the collective term for dragonflies and damselflies, is a rapidly growing area of natural history interest.

Study of aquatic invertebrates has expanded in recent years, in part because these little creatures are fascinating and quite elegant in structure and function. Biologists have learned that the numbers and types of invertebrates found in streams relate directly to water quality. Several organizations, including the Audubon Naturalist Society, now mobilize biologists and volunteers to regularly observe the invertebrates found in streams as a method of monitoring the water quality in our environment.

ROCKS AND FOSSILS

Although my focus is generally on the living parts of our ecosystem, plants and animals, I must note that there are many interesting elements of the mineral realm to study in the mid-Atlantic. Except for the Coastal Plain, where the bedrock is generally buried under deep deposits of sediments, the mid-Atlantic region is a fine place for rockhounds. Many different rocks and minerals may be found in the Piedmont, where the stress of continental collisions hundreds of millions of years ago led to the formation of a variety of metamorphic rocks. Because the pressures of metamorphism destroy all fossils in rocks, fossil seekers are advised to visit the mountainous regions where sedimentary rock layers, especially shales and coals, may contain many interesting fossils dating from eras ranging from 5 to 500 million years ago. Many fossils may also be found in the eroded material at the base of the soft cliffs that line the Chesapeake Bay's western shore in Calvert County, Maryland.

THE HUMAN FACTOR

Any examination of nature in the mid-Atlantic must account for the activities of humans. We have dramatically altered every element of each plant and animal community in the region, contaminating many environments, introducing some species into areas with often disastrous results, eliminating others from the region altogether, and disrupting ecological dynamics everywhere. We also have the opportunity to make decisions that will allow more of the natural wonders of this region to survive. The following sections of this book will return to these points repeatedly, for our effects on nature are omnipresent and usually deleterious. We still have the chance to protect the rich biodiversity of the region and to preserve many fine places. I've dedicated my life to educating others about nature because it is my belief that everyone who understands the dynamics of nature will realize that unless we protect nature and all its creatures the quality of our own lives is destined to deteriorate. It's my hope that everyone who has an interest in nature and wildlife will become an active supporter of environmental protection efforts.

The best way to learn about nature is to head outside and watch. There are many different ways to approach a study excursion into the field. The most casual approach can simply combine the joy of an outdoor walk with brief notice of the surrounding landscape, plants, and animals. The most detailed, holistic approach can become an effort to identify every plant, animal, and rock along a path, which, in turn, triggers thoughts about how all fit together in the ecosystem. Most of us follow some middle ground, as we take time to enjoy being outside, to appreciate the aesthetic beauty of our surroundings, and to try and understand some of the features and inhabitants of the landscape surrounding us. For me this usually combines a delighted recognition of things I can already identify—old friends in a way—with an effort to learn something new. Sometimes the realm of the unknown can be intimidating: It just isn't possible to learn everything! To minimize this intimidation I tend to focus primarily on one aspect of natural history for a while. First it was wildflower identification, then birding, followed by amphibians and then butterflies, and so on. Whatever my primary goal, my secondary goal is always to observe whatever catches my eye. I'll still try to key out and identify an interesting mushroom, for instance, even if my primary focus is on butterflies or frogs.

By choosing one major focus, however, I've been able to learn more quickly. This approach has given me countless highly satisfying experiences exploring the natural world.

It is with this strategy in mind that I offer the preceding section about different groups of mid-Atlantic organisms. The following sections will take a more holistic approach, looking first at our

Rhododendron with hooded warbler

region's seasonal cycles and then peeking at some especially interesting natural areas to explore. Wherever your interest lies, the keys to seeing the greatest possible variety are the same: Visit each of the major geographic regions, for you'll find a different set of habitats and communities in each region. Visit as many of the major habitats as you can within each region. Learn about and visit specialized habitats and microhabitats where species with limited occurrence may be found. Visit especially rich areas more than once per year and in different seasons. If possible, look over the shoulder of an experienced observer. Many parks, nature centers, clubs, and education organizations offer field trips and classes where enthusiasts of all levels of experience and expertise can explore nature together. If you're not a group person, books and field guides can offer a lot of assistance. Last, pick a place or two that you can visit often and watch the progression of seasons and the cycles of many plants and animals.

I hope this book will direct you to interesting places and resources for nature study. Your field studies will be more productive if you arm yourself with tools of the trade. Binoculars bring distant animals into clearer view and permit detailed study of the shape of mountainous lands. A magnifying glass or hand lens brings details of plants and insects into clear view. A small insect net wielded carefully can allow you to closely study many small creatures and then release them back into their environment. I carry a few small glass jars of various sizes in which I can place an insect for close observation. All these tools can fit into a day pack, in which I also pack a water bottle, snacks, maps, field guides, and a small notebook for taking field notes. I'll sometimes carry a small tape recorder to make verbal field notes and to record the sounds of birds, frogs, or insects. I love to hear a tape I made of periodical cicadas calling in 1987; I can't wait for their return in 2004.

Hazards of Field Study

Nature exploration in the mid-Atlantic region is not dangerous, especially if you follow common-sense precautions, although there are a few potential hazards to avoid. The sun causes more problems than anything else. In summer it's important to protect yourself with clothing or sunblock lotion. Dehydration can be a problem in this season, too; be sure to drink plenty of water. In cold weather, hypothermia, or cooling of the body's core, can be a life-threatening problem. Even in 50-degree temperatures it's important to wear plenty of warm clothing. The body chills much more rapidly when wet; high-quality rain gear is a good investment. If you can't stay warm, get out of the cold as soon as possible; don't try to tough it out. Once the body core gets thoroughly chilled and body temperature starts to drop, brain function slows and your judgment is hindered.

Animals present few hazards. Biting insects can be a nuisance from spring through fall, but they are rarely a major problem. Ticks can be common in the spring and summer. The large dog tick can transmit Rocky Mountain spotted fever, and the small deer tick is a carrier of Lyme disease. Neither disease is transmitted until a tick has spent 18 to 24 hours imbedded into a person. If you check for ticks carefully at the end of every field day and remove any that you find, your chance of getting either disease is minimal.

Three species of venomous snakes are found in our region, but none are aggressive, and snakebites are uncommon. Still, it's wise to stay aware of snakes in areas where they are most likely to occur. Expect to find a snake any time you turn over a log or stone. Watch where you place your hands and feet when scrambling over rockpiles. If you see a venomous snake, don't disturb it, tease it, or try to handle it.

Two poisonous spiders occur in our region, the black widow and the brown recluse. Although bites from either of these secretive creatures are rare, if you are bitten seek medical attention immediately. A complete recovery is likely. You should also remember that some mammals carry rabies. If a mammal acts in an unusual manner be sure to keep your distance. If you are ever bitten by a wild mammal, seek medical attention.

Poison ivy is a common plant that causes many people to suffer from severe skin irritation after contacting this plant. Less virile yet still annoying are stinging nettles, which cause a mild stinging reaction when touched. Probably the greatest danger you'll face on a nature excursion comes from highway traffic.

A Beginning Naturalist's Reference Library

I almost never head afield without carrying at least a few books, and usually I keep a small library nearby. There are hundreds of useful books about natural history to choose from, ranging from field guides and site guides to philosophical treatises. Some of the books I recommend to build the beginner's library for mid-Atlantic nature study are listed in Appendix 1. No book, though, can take the place of experience. The best way to learn about natural history is to go outside and explore. In the following chapters I hope to encourage you to do just that.

Whenever you go outside, please treat Earth and its inhabitants with respect. Don't pick plants and don't collect animals unless you're doing so for a valid scientific project. Even if your study is valid, be sure you are given permission to collect from private landowners or public land managers. Learn the rules for every public natural area you visit, follow them, and speak up when you see others breaking rules. Never trespass. Never harass wildlife. Try to leave no signs of your visit to a natural area. An orange peel or apple core may be natural and biodegradable, but each is almost as unsightly to other visitors as a discarded plastic bottle. There are many wonderful natural areas in the mid-Atlantic. We all need to actively care for them.

The Seasons

The mid-Atlantic region is part of temperate eastern North America, a section of planet Earth that has four distinct seasons each year. Each season is about three months long in the Washington-Baltimore area, unlike areas with long winters to the north and those with long summers to the south. In the following chapters I'll look at some of the characteristics of nature in each of the seasons. I don't follow the astronomical calendar, in which spring and fall begin with the equinoxes (about March 21 and September 21) and summer and winter start at the solstices (about June 21 and December 21). Instead, I follow the cyclical activities of plants and animals, which suggest that spring is better defined by the months of March through May, summer by June through August, fall by September through November, and winter by December through February.

Our seasons don't begin and end abruptly. Meteorological and biological conditions change gradually as the year progresses. Neither climate nor nature are absolutely predictable, but each season has its typical weather patterns and wildlife-activity patterns. Most of the region's residents love spring and fall because these seasons are dominated by comfortable weather. Summer's intense heat and winter's harsh cold are less universally celebrated, yet neither season is typically so extreme as to limit outdoor activity. All four seasons offer excellent opportunities for nature study, and in the following section I highlight some activities that I find most interesting. Each chapter begins with a narrative of the season's progression, followed by memories of a few favorite experiences.

2

Spring—
March through May

EARLY SPRING

Early each year I find myself eagerly awaiting spring's onset, watching hopefully for any little sign. I'm not alone in this—anticipation of spring is almost universal. For some people, spring brings an end to winter depression, with greater day length providing a cure for this darkness-induced malady. Others enjoy spring for more philosophical reasons, with this season symbolizing re-

Woodcock in night flight

birth and recovery from the dreariness of winter. I rather like the winter, but spring still has a special place in my heart. I think it's simply that this season increases the activity in our natural world, enabling naturalists to search for and observe more plants and animals. Those longer days help, too. As a primarily diurnal naturalist, I am delighted with more time to poke around outside in the daylight after work and in the early morning hours. The extra light is like a bonus on a naturalist's paycheck.

Spring officially begins with the vernal equinox, around March 21, but I always feel that the new season starts at the beginning of this month. Natural areas change rapidly throughout spring, each day bringing new growth to the plants and greater activity among the animals. Every year I try to choose a place I can visit at least once every week during spring, a spot to follow the progression of spring's lively activity. Most years I choose a place in the Piedmont Potomac Valley: Turkey Run, Great Falls, Pennyfield Lock, or Hughes Hollow. Even though I return to these places year after year, I'm still often surprised by a plant, animal, or scene I've never noticed before. Any little patch of nature you can visit repeatedly, especially during spring, offers an endless set of treasures to discover and lessons to learn.

I like to follow spring in the Potomac Valley because the season begins earlier here than in the surrounding uplands. The river's water moderates temperature extremes, resulting in a slightly warmer microclimate in early spring. I never notice this warmth directly, but its influence on the plants is obvious, as red maples, silver maples, and other trees flower a week or two earlier here than elsewhere in the region. Come April, the trees along the river will sprout leaves earlier, too.

West of Washington, D.C., where the river flows from west to east, flowers come sooner to the Maryland side. Here the slopes face south, catching more direct sun rays and acting as a natural solar collector. This is true not only of the trees but also of the herbaceous plants. Spring wildflowers such as violets, spring beauty, cut-leaf toothwort, and hepatica begin to bloom in mid-March in the warmest, sunniest spots.

The first time I systematically watched spring along the Potomac River was in 1980, when I worked near the river in McLean, Virginia. Along the river's floodplain close to my workplace was a rich forest that exploded each spring with an extraordinary wildflower display. Beginning in February I followed a soon familiar path there at least twice every week, watching the rapidly changing plant community. On my first walks I saw only the flowers of skunk cabbage in low swales, the blooms already past their peak and their thick green leaves just unfolding. In early March I found hepatica flowers growing from rock hillsides, some with purple blossoms and others pure white. Down on the floodplain I discovered that by looking under the leaf litter I could find the new leaves and buds of trout-lily, harbinger-of-spring, Dutchman's breeches, Virginia bluebells, and other species before they broke through to the light. In most years the leaves of these early wildflowers emerge in mid-March, and many bloom before month's end. But in 1980 the winter had been mild, and many blossoms opened by mid-March that year.

These flowers were the first of many that spring. The rich floodplain soils of the Potomac Valley nurture one of our region's most fertile environments. The periodic floods of the Potomac bring both organic and inorganic enrichment to the riverside environment. Thankfully, most Potomac floodplain areas in the Piedmont, which stretches from Washington, D.C. to Point of Rocks, are protected as park or nature preserve and support a fairly mature forest. Mature habitat, rich soils, and minimal human disturbance result in a highly productive environment.

As that spring progressed I watched the wildflower display grow ever more spectacular, peaking in late April when thousands of Virginia bluebells, spring beauties, and toad-shade trilliums crowded together in tight clusters. Dozens of other wildflower species were nearby, including blue cohosh, rock cress, squirrel corn, star chickweed, cutleaf toothwort, swamp buttercup, dwarf ginseng, twinleaf, Solomon's seal, false Solomon's seal, many-flowered Solomon's seal, perfoliate bellwort, wild blue phlox, and rue anemone. Each single flower was an artistic masterpiece, worthy of magnification so that I could observe all its intricate beauty. I got greedy sometimes, I must admit, and enjoyed the wildflowers on a grander canvas. I'd linger at places where dozens of species and thousands of blossoms could be seen. I haven't changed; in April I'm always heading for places with massive floral displays.

Virginia bluebells

These wildflowers, bathed in 12 hours of light daily by mid-March, are adapted to thrive in the warming soil of early spring. The vernal equinox usually occurs around March 21, but the length of daylight increases to 12 hours a few days earlier, thanks to the bending of sunlight through the atmosphere near the horizon. Much of this light reaches the forest floor in March and April, since the deciduous trees remain leafless until late April (though many will bloom before their leaves emerge). Spring's light and warmth in the forest create an ecological niche with great potential productivity. In areas where soils are rich, many plants vie for these resources, resulting in a riot of growth, layers of green that give rise to a progression of blooming wildflowers from March through May. By late June most of these species have withered back to underground storage roots, bulbs, or tubers, replaced by a set of shade-tolerant species that cover the forest floor through summer.

My primary goal on early spring field walks is to see this floral display, but inevitably I'm soon distracted. This is prime time for frog and salamander study, a big movement time for waterfowl, and the arrival season for many songbirds. Plus, if the right scrubby field can be found, a sunset may bring one of my favorite natural performances, the display flights of American woodcocks.

Many thousands of migrant birds move along the Potomac corridor, and include both species that rely on the water and ones whose habitat is the forest. During March the northward movement of waterfowl is in full swing, with common mergansers especially numerous along the river. The Seneca Rapids, between Pennyfield Lock and Violette's Lock on the C & O Canal, is an especially good spot to search for these attractive ducks—green, black, and white with long, thin orange bills. Bufflehead, ring-necked ducks, American wigeons, and many other ducks also visit the Potomac during migration, along with Canada geese and tundra swans. Wood ducks are also regularly seen in March, but instead of heading north, they settle down to nest right here in the big trees next to the river. Deepwater sections of the Potomac often attract cormorants, loons, and grebes along with ducks and geese. In March, osprey—fish-eating birds that winter in the southeastern United States—typically return to these sections of the river and to other places where fish are plentiful. Migrant waterbirds also gather at managed wetlands in the Piedmont Potomac Valley. Two of the better spots for viewing waterbirds are the Diersson Wildlife Management Area between the

C & O Canal and the Potomac just upstream from Pennyfield Lock, and the extensive pools, marshes, and swamps of the McKee-Beshers Wildlife Area, better known as Hughes Hollow.

Most of our Eastern phoebes return to the mid-Atlantic in March. I often see the first phoebe of the season as it sits low in a branch overhanging the river. Tree swallows also return in March, swooping low over the river and nearby wetlands to gobble up early emerging aquatic insects. March is also when fox sparrows are most abundant in our region. Watch for these big, rust-colored sparrows in brush piles and thickets. Rusty blackbirds can be found near swampy woods during early spring, and wet grassy fields may attract flocks of American pipits and robins. Northern mockingbirds and eastern bluebirds start to sing and establish nesting territories in March, while many species that nest later will begin their courtship singing on this month's warmer days.

Many other little signs of spring can be enjoyed as early as the beginning of March. Dandelion blossoms, generally cursed later in the year, can bring a smile as can the bursting branch tips of red maples. A single maple flower isn't very exciting, but a big maple tree with two million small blossoms is stunning.

Spring's onset seems a universal cause for celebration. Snowdrops, crocuses, forsythias, and other early flowers seem especially popular in the region's gardens. During March and April there must be many millions of bulbs blooming around the buildings in Washington, D.C.; I grow hundreds of daffodils, which bloom riotously in March. Still, my favorite flower shows are those that occur in our natural deciduous forests along the Potomac floodplain.

Tree swallows

Even though the natural world explodes with changes all during spring, April is one of the most dynamic times of the year in the mid-Atlantic. There's never enough time to observe it all. In addition to the extravagant wildflower displays, many birds, butterflies, frogs, salamanders, and other animals become very active in April, and near the end of the month our forests seem to miraculously change from gray-brown to green.

I never head out for an April walk without a wildflower guide, and I've usually stashed several other books in my pack as well. One will usually be a mushroom guide. Spring is often dry around here, but if a few proverbial April showers arrive, an interesting assortment of fungi is likely to pop up in our region's fields and forests. Many who study fungi don't simply observe and identify mushrooms, they also gather the edible ones and eat them. These fungal foragers know that the tasty and highly prized morels can only be found in late April and early May. Many morel aficionados forget about everything else at this time of year. For them, every field walk in this season has one goal—to find morels. They'll visit productive sites year after year, but don't bother asking these mushroom fanciers to tell you where to find morels; they keep the locations of productive patches a closely guarded secret. Years ago I had a morel-crazy friend who swore he could catch the scent of these oddly shaped mushrooms from a half-mile away. I don't know—maybe he could; he seemed to find five or ten for each one I found.

By mid-April we've usually seen the last freezing temperatures of spring in the mid-Atlantic lowlands. Suddenly, insects become abundant. Watch for many different kinds around wildflowers; almost all showy wildflowers are pollinated by insects. Many kinds of ants, bees, flies, and beetles become active in fields and forests during spring. Butterflies become conspicuous as the days grow warmer and longer—mourning cloak, eastern comma, question mark, zebra swallowtail, and Juvenal's duskywing can be seen in early April most years. Smaller butterflies may also be found in early spring. Only in spring can you find the mysterious little brown butterflies called elfins, which are never very common. Some of our most abundant butterflies, such as the cabbage white, clouded sulphur, tiger swallowtail, silver spotted skipper, and pearl crescent, make their first flights of the year in April. A few species occur as adults for just a few weeks each spring; search now for the falcate orange-tip, Olympia marble, silvery blue, and dusky blue or you'll have to wait until next year.

Each spring, flying insects first become numerous over bodies of water.

Because water warms and cools more slowly than does land, the danger of freezing conditions passes from aquatic habitats earlier than from terrestrial ones. Early hatches of mayflies, midges, and other such insects can result in hazy swarms over rivers, ponds, and marshes. It's no surprise that the migrant birds that feed on flying insects usually return first to habitats near water. Look for eastern phoebes, tree swallows, barn swallows, and northern rough-winged swallows to move into our region en masse during April. I always hope that the big flocks of warblers and vireos will arrive before the trees leaf out, so they'll be easier to spot, but they never do and probably never will. These birds are primarily insect-pickers, and a favorite food for many is the new crop of tender, young caterpillars, which emerge with the new crop of tender, young tree leaves. Most warblers arrive just as these leaves get big enough to feed hungry young caterpillars. Unfortunately for birders, the leaves at this time are also big enough to conceal the birds. (No wonder the recordings of warbler songs sell so well.) I am thankful that a few birds arrive early every year, and are easy to spot in the leafless trees. I always hope to find a few species by mid-April. Pine warbler, palm warbler, northern parula, yellow-throated warbler, black-and-white warbler, Louisiana waterthrush, solitary vireo, American redstart, and prothonotary warbler are generally part of the vanguard.

The early flights of insects over water in our region feed predators other than birds, too. Dragonflies and damselflies dart furiously through the air over ponds, gathering and gobbling down a variety of little critters. Anglers know that trout enjoy nothing better than a newly emerging mayfly or stonefly; to perfectly imitate a stream's insect life is the fly fisherman's goal. Frogs and toads are also guests at this vernal banquet. Many court and breed in early spring, which makes them more visible at this season. A warm, moonlit April evening is ideal frog-watching and frog-listening time. Later in the year, when frog behavior is less driven by reproductive hormones, most frogs and toads become far more wary and secretive.

Barred owls are also quite vocal during spring. Listen for their distinctive *who-cooks-for-you* hooting at twilight and throughout the nighttime hours. Try imitating the call; a curious owl may come to investigate even the calls of novice hooters. Watch closely for movement in the forest after hooting because you'll never hear the flight of an approaching owl, since their wings have specialized feathers along the leading edge to muffle sounds. A treetop owl silhouetted against a moonlit sky is an unforgettable sight.

The best thing about early spring, to me, isn't any of the birds, butterflies, salamanders, or flowers. Rather, it's the sum of them all—the momentum building up in the natural world. From spring's inception through the following half-year our world will be a riot of plants and animals: growing, blooming, expanding, changing. Almost every living thing has an opportunity to thrive, though each is challenged by potential predators and limited resources. Every year I hope to watch it all unfold anew.

LATE SPRING

The latter portion of spring means the peak of northbound songbird migration in our region, an event that for many is one of the natural highlights of the year. It's the season birders dream about all year, this time of courtship and nesting among resident birds and the peak of migration for songbirds and shorebirds. For birders there are never enough days or hours in late spring to be outside searching for birds, watching for birds, listening to birds, or just enjoying the antics of birds.

Songbirds pour through the area at this season, and every day brings a new set of species. I try to make observations wherever I am. A few years ago, while living in a rather built-up area near the Maryland-Washington line, I heard six species of warblers singing outside my window before I got out of bed. This is the season when a single tree surrounded by pavement at a shopping mall may harbor a yellow warbler, a Swainson's thrush, or some other unexpected species. Although other elements of the natural world reach a vibrant springtime peak in May, I can't stop listening for songbirds while enjoying wildflowers, butterflies, salamanders, and other natural wonders. All the natural areas in our region, from the mountains to the sea, are worth visiting in May, but daily observations of any local woodlot can also be extremely rewarding.

Shorebirds are a major May highlight. Millions of sandpipers and their allies pass through our region en route to their arctic breeding grounds. Coastal habitats along the Chesapeake Bay, its tidal tributaries, the Atlantic Ocean, and especially Delaware Bay provide feeding stops for massive flocks of shorebirds. Most birds are in full breeding plumage, which makes their identification a bit easier. The birds are a delight to see in this plumage, too: A dunlin in May, for example, is elegantly beautiful with its patterned russet back and jet-black belly. In winter a dunlin is a dull, dun-colored bird; still great to see, but not strikingly beautiful, as it is in the spring. Many other species feature rich reddish tones and natty patterns at this season.

A great shorebird phenomenon of global importance occurs along the shores of Delaware Bay each spring. Horseshoe crabs, primitive marine arthropods of truly ancient lineage, are abundant in our coastal waters. In May these animals mate, the females hauling themselves up onto the shore to lay their eggs in a great synchronous wave. Huge numbers come to the Delaware Bay shore, where coastal mud flats can be a solid undulating mass of horseshoe crabs for as far as the eye can see. It's an incredible display of natural abundance, a concentration of animal life matched by few other phenomena on this planet.

These horseshoe crab eggs are a protein-rich food source that sustains major populations of many shorebirds of the Western Hemisphere. Many shorebird species move from South America to the high Arctic each year, stopping at just a few places to refuel. Delaware Bay is the last stop for many birds, who must find enough food here to sustain their flight up to Hudson Bay, Baffin Island, or some other spot in the far north. The bay's horseshoe crab eggs give these birds enough energy to get them to their nesting grounds.

The number of birds that gather here is astounding. In mid-May, for instance, there are probably more red knots on the Delaware Bay than in all other parts of the world combined. A visit to any of the protected areas along the bay's shore, such as Bombay Hook, Little Creek, Kitts Hummock, or Broadkill Beach, can keep a shorebird fancier busy for hours, searching for twenty or more species including rare ones such as ruff or curlew sandpiper. But you don't need to identify a single species to enjoy the amazing concentrations of birds, a great writhing mass probing into the sand and mud, gobbling down the horseshoe crab eggs, and bursting into simultaneous flight with the passing of a peregrine falcon. The show reaches its peak with the extreme tides of a full or new moon in the last half of May.

Birding, like every aspect of nature study, often becomes something of a treasure hunt. It's always exciting to find something unusual, and for a few weeks in May there's the year's only chance to see and hear songbirds that pass through our region only during migration. Most can also be found during fall migration, too, but the birds aren't singing at that season. For those of us who love bird song, and depend on bird sounds to help us find and identify songbirds, spring migration is *the* time to search for these birds. Throughout the month of May, a birder has a chance to find Tennessee warbler, Cape May warbler, bobolink, yellow-bellied flycatcher, gray-cheeked thrush, and many other birds on the move. One year I'll arrange my schedule to take off work for the entire month of May. I'll surely find at least thirty species of warblers—an annual goal—and many other birds.

Every year I do take one full day to celebrate May's birds. I participate in a fundraising event for the Audubon Naturalist Society called a birdathon, where I try to find as many birds as possible in a single 24-hour period, getting sponsors to pledge a certain amount for each species that I find. Usually my total is around 150, but only in May is such a number possible. Nesting songbirds are on territory and many migrants are passing through. In May most songbirds sing vigorously, a great help for those of us with better ears than eyes. Shorebirds are abundant near the coast, especially in the areas with large numbers of horseshoe crabs. Gulls, terns, herons, and egrets are also numerous around aquatic habitats. During the 24-hour madness of a birdathon I usually manage long stops at four or five places, with brief stops at about a half-dozen more. It's maddening to try to choose which places to visit, which to skip. For the rest of the month it's one mad dash after another, trying to visit all my favorite birding spots. I spend every bit of free time in May, whether full weekends, single days off, or a few spare hours before or after work, in the field. So much is happening, and I don't want to miss anything.

Birds were on my mind when I scheduled a field trip to the Mason Neck peninsula on a May weekend one year. When I lead field trips, I always try to take the general approach to nature study, looking at many aspects of natural history, but in May I'm always thinking about birds. The month is explosively rich in other natural activity, though, and unexpected wildlife sightings are likely. On this day we were continually surprised as Mason Neck's reptiles and amphibians displayed themselves to us. Even the most ardent birders among us were constantly distracted by the *herps*, as reptiles and amphibians are collectively known.

It started as we reached the boardwalk over the swamp in Mason Neck State Park. Great thrashing and splashing sounds came from the water, where two large snapping turtles were wrestling, clutched together belly-to-belly, flipping over each other, each lunging ferociously for the other's throat. One grabbed the other's neck tightly, clamping down with its powerful jaws. Could it actually kill its rival, we wondered? More lurching and splashing ensued, and finally the two separated and disappeared quietly under the water.

Only then did we notice the brilliant yellow prothonotary warbler that had been singing all the while, just a short distance away, and the pileated woodpecker pounding on the tall trees nearby.

Before we could leave the boardwalk we were stopped again, this time by the sight of a pair of bulging eyes against the yellow face of a richly colored bullfrog in the water. We looked around and found another, then another, and finally about a dozen bullfrogs scattered around. Each one's facial color showed different tones, all more vivid than the drab green shown in the field guides. We learned that bullfrogs develop this brilliant breeding coloration for a brief period each spring. Our continued searching also yielded several northern water snakes at the water's edge, sunning on fallen logs.

Green treefrog

As we walked, we found more treasures. Little green treefrogs hopped down the trail in front of us, their lime bodies highlighted by thin golden lines through the eyes. Photo stops were obligatory. Then came a trilling call from above, sounding a bit like a sickly red-bellied woodpecker. We searched and found the singer, a gray treefrog that looked just like a bit of beech bark covered with lichens. Next we sighted a richly colored eastern hognose snake, which hissed, flattened its head, and hissed some more, hoping we'd mistake it for a rattlesnake and leave it alone. We handled the snake briefly before continuing down the trail, where we found salamanders under logs and butterflies lighting on the flowers. Every once in a while we remembered to turn back to the treetops, where the songbirds were singing, flitting, and gobbling down caterpillars. It had been a rich day afield, full of the lively and unpredictable activity that is found on almost any late spring field walk.

Though infrequent, slow days do occur in May, days when the wildlife just stays hidden. Even when I see little wildlife, I still observe plants. As my botanist friends like to point out, flowers don't run away and hide, and in late spring there are many places where wildflower displays peak. One of the best is the G. Richard Thompson Wildlife Area, located on a ridge above Linden, Virginia. Millions of large-flowered trilliums, whose brilliant white flowers fade to a soft pastel pink with age, carpet the hillsides here. Many other species can be found here as well, from the conspicuous yellow ladyslipper to the obscure but intriguing green violet. Listen for songbirds while enjoying this floral display—scarlet tanager, rose-breasted grosbeak, solitary vireo, hooded warbler, and Kentucky warbler are common spring birds here. Turn over a few rocks or logs in moist locations and you may find a red-backed, spotted, or slimy salamander. Be careful turning over objects at drier sites, where copperheads or rattlesnakes could be found. Roll over a log with your walking stick, however, and the discovery of a venomous snake won't be dangerous, though it could be startling! More likely finds include the slender little ring-necked snake or the fast-moving garter snake.

Late spring holds many other wonders in our region: Five-lined skinks dash around dry woods while spicebush, tiger, and zebra swallowtail butterflies flit around in rich floodplain woods. Trout gobble down mayflies and stoneflies in clear mountain streams. Flowers bloom everywhere, deciduous trees sprout their new green leaves, and delicate new fern fronds unfold. A hillside of wild azaleas and mountain laurels in full bloom is hard to beat. May's treasures are the memorable experiences of nature at its most vibrant.

I love spring so much that I hate to see it go. Every year I take a spring recovery expedition or two in June or July. It's not hard to do—just head someplace colder. In the mid-Atlantic's higher mountains, like those of Garrett County, Maryland, and West Virginia, spring stretches well into June. While the Potomac floodplain near Washington, D.C. mixes floral displays with new green tree leaves in late April, up in the mountains the river's headwater forests still look fairly wintry on the first of May. Some spring I'd like to follow the Potomac upstream from Point Lookout to the Fairfax Stone, trying always to be at a place where spring is at its peak.

SPRING NIGHTS

Wetlands become noisy places after dark as the courtship season for frogs begins in March. Spring peepers, our abundant little treefrogs, have incredibly loud voices. In early March there might be just a few peepers calling, their upslurred whistles bold and clear against the stillness of the night. By late in the month, however, large wetlands ring with the chorus of thousands of peepers. In the midst of such a chorus it's difficult to pinpoint a single animal, and the combined noise can be almost painful. Listen carefully, though, and a few other frogs' voices may be discerned, such as the rough bark of a wood frog, the buzz of a pickerel frog, or the nasal plunk of a green frog. These calls become more prominent in the following months, but a few are generally heard within late March's peeper cacophony. With a flashlight and a good eye you might spot some singing males, their enlarged throat pouches looking a bit like bubbles of gum.

Salamanders also begin courting and breeding in early spring, but these amphibians are silent. Bring your flashlight to any frog-filled wetland some night to peer into the shallow water; you may find spotted salamanders—beautiful, big, and black with bold yellow spots on the back—as they swim around searching for a mate. Spotteds and other species, such as the red-backed, our most abundant salamander, can sometimes be found during the day by turning over logs or rocks in the forest. Always return these objects to their original position after you've looked.

You might also hear the honking of Canada geese overhead on a March evening. Most of our wintering waterfowl begin their northward journeys in March. Geese often move at night, and they can be quite vocal. Most tundra swans will be gone by midmonth, but migration of ducks might be seen well into May.

THE COURTSHIP DISPLAY OF THE AMERICAN WOODCOCK

Peent! Suddenly our soft, idle chatter stopped and all eyes opened wide. A male woodcock had begun its preparation for a courtship flight. We looked around the field in the fading light, hoping to spot the bird on the ground. *Peent!* Binoculars up, scanning. No, can't find it. *Peent!* The calls came closer together. Finally there was a little whistling noise, and we scanned skyward. Someone spotted the woodcock, flying upward in a great circle, wings continuing to make a fluttering, whistlelike noise. Up and up went our bird, sometimes slipping out of view, until it began a parachuting descent. Its sound now was a series of twittering chirps, as the woodcock glided and dropped, glided and dropped, then did a free fall to the ground. All was quiet, and we began to breathe again. *Peent!*

Aldo Leopold aptly called the woodcock courtship display a *sky dance* in his wondrous book, *A Sand County Almanac.* I am amazed and awed every time I see this performance, and I always feel lucky afterward. I wish everyone could watch a woodcock display each year. Naturalists at several area parks and nature centers offer woodcock-watching programs and field trips in early spring. These marvelous birds don't nest near our big cities, but they are fairly common as early spring migrants, displaying lustily en route to northern nesting areas. If you're ever near a slightly overgrown open field on a March evening, be sure to stop and listen. If you hear a *Peent!*, bundle up the best you can, find a place to sit inconspicuously, cross your fingers, and watch. As the last twilight glimmer fades from the western sky, maybe you'll see a single display flight, or perhaps your bird will make ten or more flights, some to be enjoyed solely because of the bird's sound.

ONE BIRDATHON NIGHT

It was well past sunset on a cloudy night when my good friend Darryl Speicher and I turned down the dark, quiet Elliot Island Road on Maryland's Eastern Shore. We had been birding since well before dawn. We stopped several times in the pine woods, whistling for eastern screech-owls and listening for the calls of chuck-will's-widows. We continued down the road, pausing briefly at an old, dilapidated shack where our flashlight beam picked up the face of a barn owl. When we reached a vast, open marsh, mosquitoes buzzed in our ears and bit the backs of our necks and hands as we listened for bird songs over the noisy chorus of singing frogs. At one stop we heard a Virginia rail, at the next seaside sparrows and marsh wrens, and finally we detected the peculiar call of a Henslow's sparrow. We carefully noted each species on the checklist. We drove another mile through the inky blackness of this remote marshland, stopped, and heard the magical calls of black rails—tiny, secretive birds of the marsh.

The distinctive call of the black rail was all we needed to add this rare species to our list, but Darryl had never seen one. We decided to give it a try. A short distance away we heard the ringing call of another rail, *kee-kee-kurr*, sounding like it was very close to the road. We switched on a portable tape recorder, set it down on the road, and stepped back a few feet. Our recording called back to the rail as we waited and waited, slapping at mosquitoes all the while. Finally the marsh grasses at the road's edge rustled, and there it was. We trained our flashlights onto the tiny black bird; its red eyes gleamed fiercely back at us. The rail slowly walked up onto the road, passing just a few feet from us. We saw the russet patch on the bird's upper back and its finely barred underparts. To our amazement the bird flung itself against the tape recorder, backed up, and made a second attack before calmly walking back to the marsh. It chose a route directly toward Darryl, actually walking over the toe of his boot before disappearing back into the black marsh. Energized, we headed off to search for more birds of the night. When the day ended we had tallied 156 species, but we would have been happy with just the one. Black rails must be the most difficult mid-Atlantic nesting bird to see; I doubt that I'll ever see one as well again.

3

Summer— June through August

EARLY SUMMER

Summer overtakes spring during June, meteorologically at the summer solstice, around June 21. For the mid-Atlantic natural world, however, the beginning of the month signals summer's start. By the end of May most of the migrant birds have moved through the area, and our nesting species are hard

Ruby-throated hummingbird and trumpet creeper

at work, with more young hatching in June than in any other month. The early nesting season is a great time to study songbirds: Males are still defending territories, sitting in prominent spots, and singing loudly. You can still see some nest building and courtship, though most of our songbirds complete these tasks earlier in the year. Strange noises coming from the forest in June are likely to be the food-begging calls of nestlings. Have you ever heard a young great horned owl crying out at night? It's a loud, harsh noise that terrifies me for an instant, no matter how many times I hear it.

As summer begins, the springtime frenzies of bird migration and wildflower blooming end. The year will still bring many outstanding opportunities to bird and botanize, but for me, nothing matches the April profusion of wildflowers along the Potomac or May's onslaught of migrant warblers and shorebirds. So with June's arrival I recommend other nature study strategies. Either head to the mountains, where the springtime peak comes later than in the lowlands; to the coast, where summer wildlife activity is especially vibrant; or remain close to the city, making the study of butterflies or other summer attractions your major goal on nature walks. Slow your pace and watch for little details in nature.

Early summer is the time to find a comfortable spot deep in the forest, sit quietly, and wait. Early morning and late evening are the best times. Our natural world flurries with activity at this season, and any spot you choose will be a busy place. Some songbird nest will surely be nearby, though you may need to be patient to discover this. Mortality is high for eggs and nestlings, and songbirds have learned to approach their nests cautiously when potential predators are nearby. Sit still, though, and eventually your presence may be ignored. Red-eyed vireo, Acadian flycatcher, eastern wood-pewee, and ovenbird are among the birds you might find nesting in our suburban forests.

Slap! Oh yes, you might find a few mosquitoes or flies landing on you during this vigil. Energy is zipping through the food webs in our forests at a furious pace in summer, and when we visit natural areas, we become part of the ecosystem. I find some perverse satisfaction knowing that the mosquitoes biting me will lay eggs in nearby ponds and streams, and their larvae will feed little fish, frogs, and salamanders which, in turn, will feed belted kingfishers, green herons, and bigger fish. Shouldn't every naturalist become part of the food chain in this way? Of course, if you choose to use insect repellent I will understand. (By the way, did you ever wonder what to do about mosquitoes in national parks, where killing any wild animal is forbidden, but where it's also illegal to feed wildlife?)

The hot, sunny days of early summer, the longer hours of daylight, and the abundance of meadow wildflowers are ideal for butterfly studies because many species are active at this season. Butterfly study has become justifiably popular in recent years. Birders have discovered that at midday in summer, when birding is at a near standstill, butterfly watching is at its best.

Open fields and meadows in our region fill with flowers in summer, and many wetland habitats are also dominated by summer-blooming plants. Few wildflowers are as spectacular as the marshland rose mallows—wild cousins to hibiscus—or the water lilies found on quiet ponds. Lizard's-tails, common in many area wetlands, are named for the tapering cluster of white flowers they bear. Huntley Meadows, Hughes Hollow, Lilypons Water Garden, Blackwater Wildlife Refuge, and the C & O Canal are good areas to visit in search of wetland wildflowers. At any of these sites watch for turtles basking in the sunlight at the water's edge; painted turtles and red-bellied sliders are common through most of the mid-Atlantic. Northern water snakes are also frequently seen swimming through quiet ponds or resting at the water's edge, and black rat snakes often rest in the branches of shrubs and small trees next to the water. Bullfrogs and green frogs lie quietly in still pools, bulging eyes and snouts poking out from amidst the algae and duckweed at the water surface.

Two-lined and northern dusky salamanders skulk under rocks in clear streams, while brilliantly colored red efts, our most terrestrial salamanders, roam the forests. Efts darken in color and return to the water as they mature; we call them red-spotted newts as adults. Colorful dragonflies dart over fields and wetlands, while shield bugs, mantises, leafhoppers, and cicadas cling to vegetation. Everywhere you look teems with life in summer.

This is a season to enjoy at a snail's pace—it's usually too hot to hurry anyway. Wildlife abounds, but careful observation is frequently required to find much. Visit some favorite spots this month and take a folding chair. The longer you sit quietly, the more you are likely to discover. Everywhere, from the center of a wilderness to a suburban backyard, has a vibrant ecosystem to watch. Watch closely and you should find dozens, perhaps hundreds, of different organisms. Any two organisms you observe have some relationship, sometimes competitive, sometimes cooperative, sometimes a little of each. We have little understanding of the billions of interrelationships among creatures in nature. Summer is the ideal time to marvel at nature's complexity.

I have long despised the heat spells that characterize midsummer in the mid-Atlantic. We usually suffer through a few periods of extremely hot, humid days, with hazy skies that accumulate air pollutants. High temperatures hover around 100 degrees Fahrenheit during these spells. Nights are so warm and muggy that even a sheet seems like too much cover. Biting insects can be abundant. I find myself feeling hot, sweaty, and sticky all the time.

Years ago I made a pact with myself to try to escape from Washington, D.C. summers whenever possible. For five years I had great success, living and working during the summer in the mountains out west. Then, after returning to the east, I began to jealously hoard my leave each year, using every possible scrap for summer travel to those western mountains. There I found cool and comfortable weather, active wildlife, and bountiful wildflower displays.

With a limited amount of leave, however, I had to spend part of the summer back here. About the same time I had begun traveling to tropical rain forests in winter, first in Peru and then in Costa Rica. In these places I tolerated the heat and humidity in order to study the extravagant richness of tropical habitats with their tremendous biodiversity, fascinating interrelationships, and intricate food webs. These rain forests have amazingly complex ecosystems, which are outstanding places for nature study. Forest plants are specially adapted for various levels, from the canopy to understory trees, vines, shrubs, and ground covers. Insects are everywhere, feeding on leaves, pollinating flowers, and feeding the birds. Along with the exotic species, I saw familiar warblers, vireos, thrushes, and flycatchers—the Neotropical migrants.

I don't recall when the connection first became clear to me, but slowly the parallels developed in my mind: In summer, our eastern deciduous forest has a distinctly tropical flavor. Many elements of the tropical rain forest's incredible complexity could be seen right in my backyard. With this new perspective I began to observe our midsummer world more closely. I found it almost as rich as those exotic locales.

Is there a wooded patch you visit frequently? During the dead of winter, trees are leafless and signs of life are scarce in your wooded patch, but when you return to such a place in July your forest is completely transformed. Green plants take advantage of many levels of the forest, from the canopy oaks, maples, beeches, or tuliptrees down past understory dogwoods or hollies, to the shrubby viburnums or mountain laurel and on down to the ferns, clubmosses, or perennials covering the forest floor. Vines, such as trumpet creeper, wild grape, Virginia creeper, and poison ivy, take advantage of yet another niche. The stratification of our summer forest community is much like that of tropical forests.

Insects, spiders, and other arthropods are plentiful and diverse in the tropical rain forest. This is certainly true in our summer forests, too, for better and for worse. It's easy to complain about the troublesome ones, such as mosquitoes, biting flies, gnats, ticks, and stinging wasps, as well as the ones that feed on our farm and garden crops. Fortunately these pests are just a tiny proportion of our arthropod fauna. Many inconspicuous insects help keep pest populations in check, and countless species are important pollinators of both wild and cultivated plants. Flycatchers, warblers, vireos, thrushes, some hawks, and even hummingbirds feed on insects, which are an excellent protein source for the rapidly developing young of all these birds. And for a mind-boggling study of adaptation, predator-prey relationships, and myriad ecological principles, take a close look at insects.

There are many widely loved insects. Praying mantids and ladybugs (more properly called ladybird beetles) are welcomed to gardens. Fireflies (which are also beetles) are an integral part of summer's charm. Agile dragonflies, many adorned with lovely colors, fascinate many observers. Watch for mated pairs of dragonflies as they fly linked together over ponds and streams. Their immatures live an aquatic life, as do many other insects and invertebrates. Some aquatic invertebrates require extremely clean waters, whereas others are more pollution-tolerant. The relative abundance of different aquatic invertebrates is an excellent gauge of water quality. Several citizens' water-quality monitoring programs have begun in recent years; volunteers in these programs regularly assess the water quality of study areas by observing the variety, type, and numbers of aquatic invertebrates. There is also a major ecological value to these little critters—they are the primary food for many fish.

Annual cicadas, large green and black cousins to tiny leafhoppers and aphids, can be seen and heard every summer. Their buzzy calls, like the noises of all insects, are made mechanically, not vocally. The periodical cicada—which is orange-colored whereas the annual cicada is green—has an amazing life history. For sixteen years the wingless immatures, called nymphs, live underground, feeding on the roots of woody plants. During the summer of their seventeenth year these nymphs emerge in huge numbers, molt into winged adults, court, mate, lay eggs, and die. The males call raucously all day long—an incredible din. For about six weeks in summer, once every seventeen years, periodical cicadas become the most obvious animals of our forests. Around Washington, D.C., and Baltimore our next big cicada years will be 2004 and 2021. But while you wait, there are hun-

Milkweed

dreds of other interesting insects to observe or hear in summer. Crickets and katydids fill the summer night with their varied chirps and trills, and antlions, lacewings, shield bugs, June beetles, dobsonflies, carpenter bees, and many others can be great fun to watch.

Summer's insects are delightful, but so are the season's wildflowers, especially in meadow and marsh habitats. Also, fall migration for birds begins in midsummer, with shorebirds flooding our coastal areas as early as mid-July and many songbirds moving through in August. Hot summer days are perfect for reptile observation, with turtles and snakes often soaking up sunshine in plain view. Stay alert for lizards and skinks running along the forest floor, especially in dry woodlands. Midsummer is a time of rich exuberance in our natural world. In spite of the hot, hazy, and humid weather that characterizes the season, I've learned to love exploring nature at this dynamic time of year.

LATE SUMMER

Summer doesn't end abruptly—rather there's a slow transition into autumn that begins in late summer. One of our first signs of fall is the arrival of big flocks of southbound shorebirds at our coastal refuges and parks. In one of nature's more amazing adaptations, the adults of many species of shorebirds leave their nesting grounds on the arctic tundra shortly after their young have hatched. Chicks of these species are precocial—their eyes are open when they hatch, their legs become strong quickly, and their feathers grow rapidly. Because the young are able to wander around and find food without help of the parents within a few days of hatching, the older birds leave the nesting area long before the young. Ecologically this leaves more food resources for the new generation. Birders know it also brings southbound migrant shorebirds into our region as early as July. During August our coastal wetlands and other areas with substantial marshes and mud flats are filled with shorebirds, almost all adults. The young arrive later, most in September and October, with many looking quite different from their parents because juvenile plumage differs from adult plumage in many species.

Songbirds move south in August too. Have you ever followed the nesting cycle of one of our migrant songbirds, perhaps a warbler or a flycatcher? Most of these birds nest just once a year, and when the young have fledged, generally in July, it only takes a few weeks of furious feeding before their southbound journeys begin. Hawks and other birds of prey also begin to move south during late summer. Ruby-throated hummingbirds, the only hummers of the east, move south in large numbers during late summer. If you have a hummingbird feeder, or if your garden features red flowers such as bee balm, cardinal flower, or salvia, watch carefully during late summer and you're likely to see several ruby-throated hummingbirds. You won't see many of the ruby-colored throats that give the species its name, because this feature is found only on adult males, and most of the birds you're likely to see will be young.

Summer's latter stages are filled with abundant fragments of autumn. On many trees a few early red or yellow leaves appear. The world seems filled with seeds, fluffy ones blowing in the breeze, tasty ones gobbled by birds and mammals, sticky ones stuck to your socks and trousers. Monarch butterflies are heading south, sometimes in impressive steady streams, sharing the skies with southbound birds. Suddenly it seems like sunset comes awfully early in the evening, and sunrise much later. Midday heat doesn't last long, and most evenings are a little chilly. The season is changing, and winter isn't far away.

But it isn't here yet. The end of summer is still a time of diversity and abundance in our natural world. It's a season whose many treasures bring richness to the lives of curious naturalists. Some days the prize may be small: a yellow warbler in the backyard or a lovely little leafhopper in the garden.

Other days may bring the jackpot: a hundred hawks circling together or 500 monarchs in a flower-filled meadow. As the season wanes, I always try to bank the memories of summer. The season of abundance will soon be gone, and some years it seems like an eternity before spring returns.

SUMMER IN THE MOUNTAINS

In the mid-Atlantic mountains, vernal vibrancy resonates throughout June and hangs on into July and August. Wildflowers abound in the highland forests and meadows throughout the summer, with June's showing of rhododendron, rosy wild azalea, and mountain laurel particularly glorious on the high ridges of West Virginia, western Maryland, and Pennsylvania. Although the migrant songbirds zip through the mountains just as quickly as they go through our lowlands, all virtually gone by the end of May, there are many species whose northerly nesting range extends southward through our mountains, where they find habitat similar to that of New England and Canada. Nesting warblers of our mountains include mourning, Blackburnian, black-throated green, chestnut-sided, Canada, Nashville, and northern waterthrush. None of these can be regularly seen in the lowlands surrounding Washington, D.C. and Baltimore except during their migration. When it comes to warblers, mountain birding in June is almost as good as lowland birding at the peak of migration in May.

Our mountain weather is generally pleasant even through summer's worst heat waves. The scenery is great, too. Westerners often belittle our Appalachians, which can't compare in height or in craggy grandeur to the Rockies, Sierras, or Cascades, yet there is a subtle beauty to our eastern highlands. In addition to fine scenery, excellent birding, and summer-long wildflower displays, the mid-Atlantic mountains also offer great opportunities to learn about butterflies, salamanders, snakes, ferns, and limestone caverns.

My summer always includes several pilgrimages to the mountains. I'll never forget one midsummer backpacking weekend at Dolly Sods, a high plateau in West Virginia. Lowbush blueberries are common at Dolly Sods, and on this trip we caught their fruiting period at peak. My hiking buddy, Ivan Klein, and I meandered west from Bear Rocks across the open grasslands and heath barrens, eating handfuls of blueberries every few steps. Before long I decided it wasn't worth reaching for the tasty fruits unless I could grab at least a dozen in one quick sweep. My fingers, my tongue, and the inside of my mouth were all stained purple. Because of our nonstop foraging, it took us most of the day to walk the 5 miles across the plateau to Cabin Mountain. We watched a few birds, took a few photos of wildflowers, and made sure to sample a few dewberries, huckleberries, and teaberries, but we spent most of the day in a decadent blueberry feast. We'd carried food for lunch and dinner, but never touched a bit of those provisions.

We chose a small meadow near the crest of Cabin Mountain for our camp. The day had been clear, warm, and dry. We'd had no trouble with biting insects so I decided not to set up a tent, instead just rolling out my sleeping bag on the meadow grasses. Several common nighthawks passed overhead, giving their sharp calls, while the sunset colors faded and the evening star, Venus, began to shine. I crawled into my sleeping bag and cinched it closed below my chin, leaving my face uncovered. A few minutes later I was startled by a large bird that flew perhaps 6 feet above my prostrate form. I looked around and saw a great horned owl perched in a small tree at the meadow's edge. What a great ending to a wonderful day, I thought. A few minutes later the owl flew over again, and shortly thereafter it made a third pass. I began to wonder about this unusual behavior, and it occurred to me that the owl might not be connecting my bagged body to my uncovered face. My movements told the owl I was a living creature, but if it thought I was only the size of my face, might it think I was suitable prey? With visions of owl talons digging into my face, I quickly unzipped my bag, stood up, waved my arms, and hoped that I'd chased the owl away. My sleep that night was not very restful.

SUMMER NIGHTS

As summer settles into our region during June, one particular evening will arrive and I suddenly find myself filled with glee. "Firefly season is here," I'll call out with delight. These luminescent beetles, which I called *lightnin' bugs* as a child, can be quite common in all parts of our region. A few may be seen in May, but June and July are when big numbers emerge. Warm summer nights are glorious times, as fireflies fill our world with blinking magic and the emphatic sounds of owls and whip-poor-wills override the background drones of tree crickets, katydids, and toads. This is the best time to look for moths, lacewings, beetles, and other nocturnal insects around the porch light—and it's no surprise to see an insect-eating bat or nighthawk zip by.

Pesticide overuse and the decline of forest habitat in urban and suburban areas have lowered firefly populations, but early summer evenings are still well-decorated by their blinking lights. Attracting fireflies is great fun: Cover a flashlight with aluminum foil, poke a small (firefly-sized) hole through the foil, and blink the light in a mimicry of the firefly blinks. Careful blinking sometimes brings a curious firefly right onto the light.

I've read about a firefly display along some southeastern Asian rivers where thousands of luminescent beetles blink synchronously. Imagine slowly floating downstream through the rain forest along a wilderness river, rounding a bend, and seeing 20,000 little lights blinking on and off together! Our fireflies present a much different spectacle, a bit more subtle but still spectacular. Watch carefully if you find a good collection of fireflies some evening. We have several different species in our region, and each can be identified by its blink pattern. The first species seen in early summer flashes with a quick repetition of short blinks. Flightless females live on the ground, attracting males with their species' distinctive blinks. Some females occasionally mimic the pattern of another species, attracting dinner instead of a mate. Life's not always easy for undiscerning male beetles.

Fireflies are just a part of the charm of summer nights. One June evening some colleagues and I had lingered until sunset at our workplace in Chevy Chase, expressly to watch the firefly show. As the last hints of sunlight disappeared into the western horizon and we all enjoyed the firefly light show, I idly started to whistle like an eastern screech-owl. Before long there came a responding call, and soon I took the foil off my flash-

Bats at twilight

light and its full beam caught the little owl in a nearby tree, its eyes wide open in what seemed like an irate stare. It flew away after a minute or so, but the night was not still. American toads trilled back in the woods, and crickets chirped nearby. Once or twice I heard the high squeak of a flying squirrel from nearby trees. Bats darted after insects overhead while moths that had managed to evade the bats swarmed around the porch light of the nearby building. Foxes, raccoons, opossums, and other mammals were no doubt nearby, though we didn't see or hear them that evening. The air felt warm through the evening hours, and rich, earthy scents drifted through the forest. It made me want to keep watching and listening right up to sunrise.

BUTTERFLY STUDY

During summer a patch of milkweeds and other summer wildflowers may attract many showy butterflies including the tiger swallowtail, monarch, great spangled fritillary, and the coral hairstreak. A muddy spot on a sun-dappled forest trail may attract little gems such as the pearl crescent, checkered skipper, and the eastern tailed blue, while red-spotted purples, little wood satyrs, and hackberry butterflies patrol nearby. Watch carefully and you may see a butterfly's long, coiled proboscis unwind as it probes into a flower for nectar. If you spy the flight of a butterfly that looks heavy and awkward, try to get a closer look; it may be a mated pair flying together with abdomens joined. Gently net a butterfly (make sure netting is permitted in the areas you visit) for a closer look; flat-tipped tweezers, like those used by stamp collectors, are ideal for handling butterflies. Check the detail of the scaled wings, eyes, or other parts with a hand lens. Look for evidence of bird strikes. Birds eat many butterflies, and often a butterfly will be found with a flycatcher bill-shaped piece missing from a wing.

Shift the focus from adult butterflies to immature stages and the treasure hunt becomes almost limitless. Caterpillars can be hard to find and identify, but they have a wondrous variety of colors, patterns, and adornments. If you find an abundance of caterpillars of a known species you can take one or two home, provide ample supplies of the proper host plant as food, and with luck witness the miracle of metamorphosis. A butterfly pupa is called a chrysalis, and most are well-camouflaged and extremely hard to find in nature. Only with a captive critter or great luck in the wild will you have a chance to watch as a caterpillar molts into a seemingly lifeless chrysalis, which splits open a week or two later as an adult butterfly emerges. This transformation occurs millions of times a day on this planet, as it has for millions of years, yet to me it is still miraculous.

4

Autumn—
September through November

EARLY AUTUMN

I enjoy fall's weather more than that of any other season, especially its trademark breezy, crisp, blue-sky days. It's finally cool enough to enjoy brisk activities, such as a long hike or bike ride, without becoming uncomfortably hot and flirting with dehydration. Nature highlights of this season include fall bird migration, dramatic foliage displays, and opportunities to watch plants and animals in transition. If organisms are to survive the upcoming winter, they need to change their behaviors and physiological adaptations from those of summer. Fall is the time to observe these changes.

Snow geese in flight

Although autumnal events begin in midsummer, it is usually during early September that weather and wildlife activity suggest that the season has truly changed. Large numbers of birds are moving south. Mammals feed furiously in preparation for winter. Days grow shorter, many grasses and herbs go to seed and turn brown, and the first changing leaves of deciduous trees paint bits of bright red or yellow onto the landscape. Summer doesn't give up easily, though, and until the first few freezing nights many flowers and insects can still be found. Nights are downright noisy with the stridulations of tree crickets, katydids, and other Orthopterans. The fecund riot of life that characterizes summer continues through early fall, especially in the warmest part of our region, the Coastal Plain. Thanks to the moderating influence on temperatures by the sea, freezes generally come last to those areas near the coast.

Old fields and managed meadows are at their best in early fall. Plants of the composite family abound in these habitats, and many are in bloom at this season. One composite, joe-pye weed, especially amazes me because it often grows to 10 feet between May and August, only to die back to its roots with the autumn frost. By September joe-pye's big round clusters of pink flowers are both beautiful and busy. Insects love these flowers, and a variety of wasps, beetles, bugs, and butterflies are usually found amidst the blossoms. Among the butterflies, two little skippers, the sachem and the fiery skipper, are likely to be found. These two butterflies cannot survive our cold winters in any of their four life stages; one finds them only in the Deep South as winter gives way to spring. But every year sees several generations, a rapid population growth, and a mass movement of millions of butterflies hundreds of miles northward into territory that's only temporarily hospitable. I marvel at the strength and spunk of these skippers as I watch them in late summer. Their genetic lines are certain to fail, with only their cousins in the south likely to contribute to

Joe-pye
weed with
butterflies

next year's generations. But individuals from these species will be among the first to colonize new areas as climates change, so from an ecological perspective, there is a reason for this apparent genetic suicide.

Although monarch butterflies can't live through a mid-Atlantic winter, they stream north into our region every summer. No genetic suicide here, however, for the movement of monarchs is directional, purposeful, and cyclical, so we call it a migration. Rivers of monarchs can be seen flowing south over the coast of the Atlantic Ocean, the Chesapeake Bay, and above mountain ridges throughout the mid-Atlantic during early fall. Even though many monarchs fail, some survive the long migratory flight to a small region in the Mexican highlands. Here monarchs spend the winter rather quietly in astounding aggregations, millions strong, before heading north again in spring. Three to five generations cycle in our area during the summer months before the next year's southbound migration begins, yet those butterflies head back to the same winter home of their great-grandparents. How they know to do this is one of nature's many great mysteries.

The migration of birds is a wondrous mystery, too. Early fall is the time to search for southbound songbirds in the woods and field edges, for shorebirds in the wetland wildlife refuges and along the coast, and for raptors overhead wherever there's a clear view of the sky. Broad-winged hawks are the only raptors that regularly migrate in groups through our region, and most pass overhead in late September. Every time I step outside at this season I glance skyward, and once in a while I'm rewarded with the sight of three, eight, or sometimes two dozen broad-wingeds overhead. Fall hawk watching isn't easy, though, as the birds are often extremely high in the sky and hard to observe. I've pointed skyward excitedly more than once at an autumn raptor as a friend scanned, struggled, and finally spotted the speck in the sky. "*That's what you've been watching?!*" comes the exasperated cry.

Most reptiles spend the winter in a chilled, dormant state but on warm, sunny days in early fall they all seem to be trying to gather and store up as much warmth as possible, as if they had solar batteries. Visit any little pond or stream on one of these Indian summer days and you're sure to find basking snakes and turtles—try Huntley Meadows or Mason Neck in Virginia and Blackwater National Wildlife Refuge or the C & O Canal in Maryland. Painted turtles and northern water snakes will certainly be abundant, but you may also find red-bellied sliders, snapping turtles, corn snakes, hognose snakes, and other species.

The progression of fall accelerates during October. Raptor migration picks up momentum this month on mountain ridges. Thousands of birds of prey may pass over any given spot on any October day, each heading south to its wintering grounds. Snakes and turtles can still be seen soaking up the sun's warmth, while warblers and thrushes slip quietly through the woods. Late-season flowers, such as tickseed sunflowers, are still visited by a handful of butterflies. Yet even though wildflowers, beetles, butterflies, reptiles, swallows, and flycatchers are still around, it won't be long before each disappears from sight, for late autumn brings dramatic change to mid-Atlantic environments.

LATE AUTUMN

Autumn in the mid-Atlantic is sharply divided; the first part is much like late summer, just a little subdued. The second part is harsh and wintry. The transition usually occurs dramatically in late October or early November, with the arrival of winter's first monster cold front. Our region's winter is dominated by cold that builds over Canada and pushes into the region from the northwest. Summer systems push warm, moist air from the Gulf of Mexico into the mid-Atlantic on southwesterly winds. Spring and fall show a mix of these systems—a few fresh northwesterly fronts bring pleasant days in summer and early fall, but when late October arrives the watch begins for that first monster front, the one that barrels through the area on 30–40 miles per hour winds, dropping temperatures 20 or 30 degrees in less than half a day. Such fronts are the exclamation points in our emphatic autumn season.

Many years ago I experienced one of these fronts as I hiked up Robertson Mountain in Shenandoah National Park. One of the year's first big cold fronts rolled in with cold rain and strong, gusty winds. Leaves, some yellow and red, others brown, blew off the tree branches by the millions, swirling around at my feet and filling the sky overhead. Everywhere I looked, I saw ruby-crowned and golden-crowned kinglets clinging to the low branches of trees, bobbing with them in the wind. Since a kinglet weighs only 6 grams, it was clearly a struggle for each bird to hang on. From my mountaintop overlook I could see the landscape change as harsh winter conditions overtook the last remnants of summer. Big patches of forest became leafless while I watched, and ravens rocketed by, seeming to delight in the change of seasons. Although I got a bit chilly and wet that day, I felt privileged to witness such a dramatic scene.

I always watch for these cold fronts because I love their drama. Sultry, hazy, still air generally precedes such a front, followed by a sudden line of gray clouds that rolls in from the west, finally blasting across with a strong, chilly wind. Sometimes there's a rain shower, at times a little thunder and lightning, but always there's the wind. The front usually passes quickly, but the wind blows on, often for days. I think of our year as having two major seasons, hot (summer) and cold (winter), with long transition periods in-between. That first monster cold front turns the page from summer to winter; the rules for life in our natural world have changed. It's time to learn the winter scheme again.

Whipping winds and crunching of leaves underfoot provide the rhythm of late fall. On the days when the wind blows hard and cold, it sends brown leaves flying, makes tree trunks groan, and whistles through the newly barren branches. As leaves pile up on the ground, a foot thick in rich forests and deeper in the drifts, walking becomes musical. Every step results in a crackling noise; it's hard to tiptoe through the autumn woods. I've cursed the falling leaves and the wind more than once when trying to rake this litter to create a tidy space around my house. I have also rejoiced, like the time I worked all summer in the west and drove home in the fall, camping all the way. My first eastern camp was in the Ozarks, and my bed that night was a 3-foot pile of leaves that I had gathered. I nestled into that pile for a wonderful sleep. Home at last!

For insects the change from early to late fall means high mortality. Most insects simply succumb to freezing weather, though all have adaptations that allow at least a small percentage to survive the winter. With some species only the eggs are winter-hardy and able to withstand freezing temperatures. Others may overwinter in a dormant condition as larva, pupa, or adult, generally in a sheltered place such as underground, in tree cavities, under the leaf litter, or perhaps in your basement. Late fall is when you're most likely to find crickets and other insects crawling into your home, seeking its warmth.

Many insects adapted to aquatic lifestyles can stay active through winter. Water retains heat far longer than does land; even when the surfaces of our ponds and streams freeze, the temperature of the water and mud at the bottom stays above the freezing point. Dragonflies, mayflies, caddisflies, stoneflies, and several other groups have terrestrial adults and aquatic larvae; winter survival of the aquatic forms is one advantage of this adaptation.

A few insects actually migrate. Monarch butterflies are the best-known example, but a few other butterfly species head for warmer climes. Watch for southbound buckeyes and cloudless sulphurs in early fall, especially along the coast. Some dragonflies also migrate long distances in spring and fall. Perhaps some lesser known, overlooked insects also migrate; we know little about the life history and ecology of most invertebrate animals.

Regardless of the strategy employed, few insect species have high winter survivability. Whether taken by the elements or by predators (watch brown creepers pick insect eggs from the crevices in tree bark), these insects rely on high reproductive rates that allow them to resaturate their environments in a few quick generations the following summer. Herein lies a primary difference between the ecology of the temperate zone and that of the tropics. Our insect populations fluctuate wildly across the year, and insect predators must cope with these fluctuations. In most tropical ecosystems such climatic regulators don't exist, and each population is controlled primarily by the other organisms in its ecosystem.

Brown creeper

For many birds the key to surviving the cold weather and limited resources of late fall and winter is to go elsewhere. Like other warm-blooded animals, birds can survive very cold temperatures as long as food is plentiful. Heat is a by-product of metabolism. Birds that can find plenty of food will generate significant heat and hold this warmth thanks to their natural insulation, built-in down jackets. But for many birds, especially those that feed on flying insects, the food supply disappears in winter, and to survive they must migrate. Some of these migrants take their cues from the shorter day length in mid-summer and begin their journeys then. By October many warblers and shorebirds have already arrived in the tropics. For others, including most raptors and waterfowl, the months of October and November bring the greatest numbers of migrants into and through our region. Some species that nest in the far north, such as rough-legged hawk and brant, will stay until spring. The mid-Atlantic winter, with snow rarely covering the ground for more than a few days at a time, has food in good supply all winter for many sparrows, finches, yellow-rumped warblers, and other songbirds. Large bodies of water, including lakes, rivers, bays, and the ocean, can also support much bird life throughout the colder time that begins in late autumn.

Temperate-zone plants must also be prepared for the freezing weather that comes in late fall. Most deciduous trees drop leaves and become dormant by early November, after their brilliant fall colors have faded. Watch a local forest patch regularly during late fall to witness this dramatic transition from leafy to leafless. Note the branching patterns of the deciduous trees; these patterns become obvious only after leaves have dropped. Many trees produce seeds in the fall; see how many you can find. Look for buds on tree twigs; next spring these buds will open into leaves or flowers. Conifers continue to photosynthesize through the coldest months, their slender needles adapted to minimize moisture loss. Freezing weather locks up water supplies, so desiccation is a major problem for plants in winter. Many herbaceous plants simply die back to the ground, ready to burst back into glory come spring. Seeds are abundant in our fall fields and forests, most remaining dormant until the soil warms and the days lengthen in March and April. Our mild winters allow some plants to stay active through the coldest months. Wintergreen, Christmas fern, and winter cress are a few of the plants whose names refer to their green and active winter condition.

The fallen leaves of late autumn play an important ecological role, adding organic matter to the forest soils. This natural compost improves the soil's texture and creates an energy reservoir for all forest plants to draw upon. Mushrooms, often abundant in our fall forests, are nourished primarily by decaying leaves and woody twigs and logs. In years when autumn rainfall totals are average or better, the season is likely to produce a great assortment of mushrooms. Look for several species of shelf fungi on fallen logs and standing snags, and for gilled, pored, or toothed fungi growing from rich forest soils. Grab a handful of forest soil and smell its richness; part of the earthy scent comes from fungi and other decomposers—the organisms that convert organic material into soil.

Some years our oaks produce a big crop of acorns, and in these years gray squirrels have plenty of food and enjoy great reproductive success. In years when the seed crop isn't so good, the squirrel population is stressed; many weaker animals starve or are chased from their home ranges, and a good number are struck by vehicles. Others are eaten by predators. Like many other animals whose populations naturally cycle up and down, squirrel populations rise after abundant nut years, and drop after poor ones, but even when the population experiences high mortality the strongest animals survive and have modest reproductive success.

Raptors are still migrating in late autumn, with November the best time to search for two of our least common species, golden eagles and Northern goshawks. Red-tailed hawks are plentiful along mountain ridges in late fall, especially when strong northwest winds are blowing. None of the raptors migrate at night; instead, each stops to roost wherever it ends its day's migratory flight. Early and late in the day, the behavior and cries of crows can often help you find a roosting raptor because when crows find a perched hawk they'll gang up and scream incessantly, swooping down on the beleaguered bird until they chase it from their territory. Crows are often noisy, but when they've cornered a raptor their calls are different, somehow more strident and angry. Crows are generally quite vocal in fall when they congregate in huge nighttime roosts and squabble constantly to establish dominance. One of the region's largest crow roosts occurs in a small forest patch next to Montrose Road in Rockville, Maryland—an estimated 100,000-plus birds congregate here each night from late fall through early spring.

The mid-Atlantic mountains have many wonders to explore during the colder months. Shenandoah National Park empties out after October's foliage show, yet the viewpoints along the Blue Ridge continue to offer a gently colorful mosaic of earth tones. November's cold fronts bring intensely clear air into our region, pushing away the polluted haze that hangs over the Blue Ridge during the warmer months. Views across the Shenandoah Valley, itself a beautiful patchwork quilt of farms and forests, reveal several ridges of the high Appalachians. On breezy, clear and cold November days the vistas often include migrant hawks and vultures. The park's resident ravens are conspicuous on these days, screaming defiantly as they sail overhead.

Down in Shenandoah National Park's forest, November brings witch hazel, the year's last wildflower, into bloom. I'd probably never notice this shrub's skinny little yellow petals if they bloomed in another season, but borne on leafless branches in this improbable season, witch hazel always delights me. Newly fallen leaves cover Shenandoah's trails in November; it's great fun to hear them crunching underfoot and to smell their earthy fragrance. I force myself to stop walking from time to time so I can listen to the forest. Often the woods are quiet, the only sounds the sweep of wind through the barren treetops and the swirl of leaves on the ground. Songbird migration is over by late fall, but some birds spend the entire winter in these woods, generally traveling in flocks of mixed species. Once in a while I'll stumble across such a flock, Carolina chickadee and tufted titmouse being the most vocal. Ruby-crowned kinglet, hermit thrush, purple finch, yellow-bellied sapsucker, and many other species can sometimes be found in these groups.

Another sound sometimes punctuates Shenandoah's November quiet—the ringing blast of a gunshot from the valley. November is usually deer-hunting season, and casual naturalists need to be aware of the activity of hunters. Because I have no interest in walking through a forest where hunters are active, my autumn forays to the mountains are always to places such as Shenandoah National Park, where hunting is prohibited.

I'm fond of the scenic landscapes of late autumn. Brown, tan, and gray are the season's colors. November's subtle tones may disappoint some people, coming as they do after a series of vivid scenes. The colorful floral displays of spring and summer, followed by October's extravagant foliage displays, are never equaled by November scenes. Still, I love late autumn's gently soothing palette of earth tones. As the colors fade, temperatures fall, and the daily hours of sunlight drop as the month progresses. This combination seems relaxing to me—it triggers my hibernation instinct. Thankfully this season has a complex, multi-tonal alarm clock. It combines the honking of snow geese over a coastal marsh, the cries of ravens over the Blue Ridge, and the crunching of leaves underfoot. I always hope to catch every subtle note of late autumn's melody and harmony, to notice every shade in its earthy hues.

Fall begins as a time of abundant, obvious natural transition, but by the season's end our winter patterns have settled into place. Still, as witch hazel blossoms linger on branch tips, the spicebush and red maple buds are beginning to swell. There's never a time when, upon careful inspection, we can't find something in our natural world illustrating the constant seasonal flow of our temperate ecosystem. It just requires a little closer look at the time when fall gives way to winter.

It's a different world out there after dark. Not really, of course. There's no magical transformation in which plants, animals, rocks, and landscape features disappear and reappear. It's just that everything looks different in the darkness of night (which is never completely black, even far from the city when there is no moon above). One's depth perception suffers and color vision disappears; the world seems flat, all silhouettes of black and murky gray. Lacking our usual visual command of the landscape, most of us feel a bit uncomfortable outside at night. We can't see what we're stepping on, we can't see where we're going too well, we can't see what might be watching us. With time, most of us conquer some of these apprehensions and learn to love the mystery and freshness that nighttime brings to even our most familiar settings. A little moonlight certainly helps. Autumn nights, especially when the moon is full, are among my favorites.

Our world makes very different noises after dark. In early fall there's a rich chorus of Orthopterans—insects of the grasshopper, cricket, and katydid guild—that's a perpetual background sound from dusk until dawn. Sometime during July the chorus begins; I'm always delighted that first time each year when I hear the droning sound of snowy tree crickets and their relatives as a universal background whine. It evokes memories of youth, of the times when I went through the transition from fearing the woods at night to loving it. Even the familiar field cricket's chirps evoked mystery, as I learned the folklore that by timing their calls you could tell the temperature. Alas, I could never remember the formula. Later I learned to listen for the staccato bursts of three to six unmusical notes made by katydids, whose sound delights me to this day.

Other sounds enrich the nocturnal concert throughout the transition from summer to winter. Owls can be heard calling, with off-key youngsters trying to learn the right patterns, pitch, and tone. Sometimes I'll call back; the young birds are often curious and will respond or fly closer to me. American toads still offer their trills to the night, and a few spring peepers sing in early autumn too, perhaps mixing up the photoperiod signals of the two equinoxes. Deer will snort, foxes bark, yip, and howl, and if the weather's been dry, every little vole and mouse that scurries through the forest add a mysterious rustling sound to the symphony as it crunches the dried leaves. More than once I've taken a portable tape recorder outside to make a collection of sounds. I like to play these tapes back from time to time, especially in midwinter.

By the time autumn has changed from its summerlike early stages to its wintry end, the sounds of the nocturnal world have dramatically changed. Whereas early fall nights ring with the sounds of summer's chords, freezing nights quiet the insects, leaving a spooky silence just waiting to be broken by the boisterous honking of migratory Canada geese, one of my favorite sounds.

Katydid

Once or twice every fall I'll hear this marvelous music some evening when it's least expected, sometimes right over my house in the sub- urbs. I'll race outside and peer skyward, rarely with any success.

Perhaps the most poignant sound of late autumn is the intense quiet of the night. Once frosts have silenced the singing insects, nights can be remarkably silent. I love to walk in the moonlight of late fall and winter. In the deep stillness of such a night, the sudden cry of a barred owl can be very startling. It's on these nights that the rustle of animals among the dry leaves of the for- est floor seems extraordinarily amplified. Mice sound like raccoons, and the 'coons sound like bears. Deer sound like, well, my imagination used to come up with a different monster each time I heard them on a winter's night. Then, when foxes would cry out—sounding like the victims of a horror- movie villain—I'd surely want to head home and double-lock all the doors.

Like many good symphonies, the dramatic music is complemented by the peaceful. Calls of eastern screech-owls and great horned owls are pleasant and calming. The gentle squeaks of flying squirrels, masked by insect noise for much of the year, can sometimes be detected in autumn. The pleasant gurgle of streams or the sound of waves on the bank of a river or lake, carry much farther into the forest when the trees are leafless.

HAWK WATCHING DURING FALL

Hawk watching is one of the most addictive natural history pastimes. I know many passionate observers of these avian predators who spend every autumn weekend searching for hawks. Some hoard their year's leave so they can take long autumn vacations to visit places where large numbers of raptors occur. My own hawk-watching addiction is fairly mild but has passed through many stages. First came my introduction to hawk watching, when friends coaxed me to spend a weekend up at Hawk Mountain, Pennsylvania. Here on Kittatinny Ridge, the easternmost ridge in this section of the Alleghenies, great numbers of hawks and their allies move south, thanks to updrafts caused by the deflection of northwest winds by the mountains. A dip in the ridge brings the birds lower, where from the Hawk Mountain overlooks you may see them passing nearby. On that first weekend of my hawk-watching career I sat on the rocks of the North Lookout amazed as sharp-shinned hawks steadily streamed by, some swooping low, others so high as to be barely visible. Chance had brought me to Hawk Mountain for a record day; never before or since have as many sharp-shinneds been observed in a single day.

Fourteen species of raptors regularly migrate through our region during the fall. There are three accipiters—sharp-shinned hawk, Cooper's hawk, and goshawk; two eagles—bald and golden; the osprey; one harrier (northern); three falcons—peregrine, merlin, and American kestrel; and four buteos (or soaring hawks)—red-tailed, red-shouldered, broad-winged, and rough-legged. On my first visit to Hawk Mountain I saw thirteen of these, several as new life birds, and experienced a record day. Is it any wonder that I was hooked?

Only once did I make the pilgrimage to Hawk Mountain in September; I found sunny, warm conditions on the North Lookout that welcomed passersby and casual observers. Over the distant ridges and valleys I watched groups of broad-winged hawks circle upward together in columns of rising air, known as thermals. Other species also migrate in September; on my September trip I frequently sighted osprey, red-shouldered hawk, American kestrel, and sharp-shinned hawk.

Curiously, I've never made a second September visit to Hawk Mountain. I have a fixed image of the perfect day at this place set in my mind, one of ominous gray clouds and a cold, cold wind pushing in from the northwest. On such a day the regulars come with wool pants and caps, gloves, long underwear, down jackets, insulated seating pads, and vacuum bottles filled with hot drinks. Fog and drizzle may cover the mountain once in a while, stopping the flights, but as each squall passes the hawks return. The uninitiated come underprepared for the cold and don't last long. Red-tailed hawks glide by, and everyone hopes for a goshawk or golden eagle. Blasting by on the strong winds are dozens of sharp-shinned hawks, with a bigger Cooper's hawk flying by for every five or ten sharpies. (To match the regional dialect, influenced by the Pennsylvania Dutch culture, I learned to pronounce the double o in Cooper's as in the word cook, not loop.) September days at the lookout are never like this—a good thing, most would argue. I'll have to block out my icy ideal image and try September on Hawk Mountain again.

Unless you're especially fond of the history or mystique of the place, there's really no need to travel all the way to Hawk Mountain to enjoy the raptor migration in autumn. Many excellent vantage points are closer to home, including mountain overlooks such as Waggoner's Gap in Pennsylvania, Monument Knob in Maryland, or Snickers Gap in Virginia. Coastal areas, as mentioned earlier, are important raptor migration corridors, with the Chesapeake Bay shore often hosting good flights—try Fort Smallwood or Sandy Point along the bay's western shore. Chincoteague National Wildlife Refuge is a regular stop on my autumn schedule, and now I spend a little time hawk watching on every autumn Chincoteague trip. The causeway that runs between Swan Cove and Tom's Cove, just before the road reaches the beach parking lots, is a prime spot. I'll sit there on a sunny midday scanning northward with my binoculars, picking out specks in the sky as they appear over the distant trees. Cape May, Cape Charles, Point Lookout, and other spots on the Atlantic and Chesapeake coastlines also offer hawk-watching rewards.

Although many birds concentrate along the ridgeline and coastal airways, some raptors pass overhead almost anywhere in the region. Every time I spot a raptor or some other interesting bird, which happens surprisingly often, it feels like a bonus. As a result, hawk watching is now integrated into my daily autumn activity. Even though I rarely get back to Hawk Mountain, or even to Waggoner's Gap or Washington Monument, and I may not see the great concentrations or share the camaraderie of the Lookout, I still see many raptors. I see them as part of the big picture, passing overhead while I'm watching a five-lined skink scurry up the rocks at Great Falls, or while I'm following a yellow-throated vireo through the trees. Early autumn is a great season for enjoying scenes like these, as songbirds such as the vireo use all our forested areas as migratory stopovers, and reptiles such as the skink bask in the sunlight of the year's last really warm days.

THE COAST IN AUTUMN

Early fall is a great time to visit the coast. A great motivation for me is the beach cleanup, an annual event scheduled on a Saturday in September for many of our region's beaches. I like to help with the cleanup on the southern end of Assateague Island. It's amazing how much fun people have during this event, which at first sounds very much like work. The key is in the numbers: There are enough volunteers so that each person's task is manageable, and yet all of us have the satisfaction of being part of a major project. Data on the type of trash collected are gathered and later compiled; we discover that much beachfront trash has washed ashore from elsewhere, usually from boats. Volunteers have an informal contest to find the most unusual trash. Each year a large number of rubber tires are collected, which I found odd until someone explained that, years ago, the town of Ocean City strung together thousands of old tires and sunk them offshore, hoping they'd function as an artificial reef to protect the bathing beaches. The sea eroded the connections, freeing the tires to move southward on the longshore current and wash up on the wild Assateague Beach.

Even though beach cleanup participants enjoy the tangible accomplishment of the tidy beach they leave behind, there are certainly secondary bonuses of wildlife-viewing opportunities. September is a lively time on our coast, and there's always something to observe. Bird migration is definitely a highlight. Shorebirds are often pouring through, with sanderlings, willets, and whimbrels often seen along the beach and many other species on nearby mud flats and marshes. Many raptors move south over the coast, with falcons especially common in late September and early October. Often all three of our region's falcons—the kestrel, merlin, and peregrine—can be seen in a single early fall day from any of our Atlantic beaches. Gulls, terns, pelicans, herons, egrets, and many other conspicuous birds are abundant in this season. Be sure to walk through the maritime forest or coastal shrub community, where many migrant songbirds can often be found. In addition, many seaside plants bloom in early fall; others are especially lovely as they go to seed. Sea lavender, thoroughwort, and seaside goldenrod are at their showiest during this season. Monarchs, cloudless sulphurs, and buckeyes are often common around these flower patches.

WATERFOWL

One of the major highlights of autumn in the mid-Atlantic is the return of migrant waterfowl from their nesting areas to our north. Large numbers of ducks, geese, and swans spend the coldest months of the year around the lakes and bays of our Coastal Plain. Canada geese are the most plentiful species and are seen by mid-October around almost any body of water larger than a pond. Around the Chesapeake Bay the flocks often number more than a thousand geese. Although some Canada geese have stopped migrating in recent years, having learned that mid-Atlantic parks and golf course ponds can be suitable breeding areas, many more are still migratory. In places these nonmigratory geese are now so numerous that they are considered pests, but I still celebrate the arrival of the fall migrants, hoping to hear a big flock on each autumn night. It's easy to tell the difference between these wary, wily winter visitors and our tame, begging residents.

As the geese arrive in midautumn, they are joined by a host of other migratory waterbirds. Ducks, most of which have recently molted into their bright winter breeding plumage, flood into our rivers, lakes, bays, and marshes. It's always a joy to find big flocks of ruddy ducks, canvasbacks, and lesser scaup in the tidal Potomac. Scoters migrate just off the ocean beaches, seeming to work like crazy just to stay a few feet above the ocean surface as they fly south. Double-crested cormorants also move south following the coastline, flying in imperfect lines and vees. Loons and grebes arrive too, sometimes stopping in freshwater lakes and big rivers, but generally continuing to the brackish waters of the Chesapeake or the open ocean. Every few years an ice storm occurs in November, and some migrating loons or grebes mistake an icy parking lot for a lake. They land on the parking lot at night and cannot take flight again because they need to run across the water's surface, wings flapping wildly, before they can become airborne. Great rescue efforts are mounted after these storms, with volunteers picking up stranded birds (cardboard boxes make excellent holding cages) and driving them to nearby lakes and rivers. A horned grebe once spent some time in a bathtub at my workplace, its midpoint between rescue and release. It seemed surprisingly content there.

Tundra swans and snow geese arrive in late fall, too, and each adds its voice to the seasonal concert. Our coldest season must not seem too bad to the swans, which nest in the arctic tundra, as their name implies. The Chesapeake Bay and its associated marshes are a major wintering area for these huge, majestic birds. Drive the roads through Eastern Shore farm country between November and February and you're certain to see flocks of swans, along with larger groups of Canada and snow geese, standing in the farm fields. These birds have learned that old fields still bear fruit, as mechanized harvesters leave a small percentage of grains in the fields along with the stubble. Visit one of the marshy national wildlife refuges on the Eastern Shore, such as Blackwater, Bombay Hook, or Eastern Neck, and you're likely to see hundreds of swans and thousands of geese. Part of the management plan for these refuges is to lease land to local farmers, who till, plant, and harvest fields of soybeans and corn; geese and swans feast on the leftover grain. Snow geese congregate in huge flocks; sometimes you may see 10,000 or more in the sky. Listen carefully to the honks of snows—their call is a little higher and lighter than the call of Canadas.

Although most snow geese remain east of the Chesapeake Bay, tundra swans can be found most winters in many locations closer to Washington, D.C., including the tidal Potomac, off Alexandria, and Little Seneca Lake in Black Hill Regional Park. Any farm field or pond might attract a few swans in November, though it's always a surprise to see them. Canada geese may also be seen closer to home, both the year-round residents and the migrants. Find water in the autumn and you're bound to find Canadas. Merkle Wildlife Sanctuary, just south of Jug Bay on the west bank of the Patuxent River, supports huge wintering flocks of Canada geese, along with much other wildlife. Ducks are also found in most water habitats during late fall. The Piedmont Potomac, especially around Violette's Lock, is a fine place for common mergansers, as is the lower Susquehanna. Near the Chesapeake Bay Bridge, as viewed from Sandy Point State Park, swim oldsquaws, common goldeneyes, and bufflehead. Green-winged teal, American wigeon, northern shoveler, and many other dabblers visit the Great Marsh at Mason Neck National Wildlife Refuge. I never pass a marsh, pond, lake, or quiet stretch of river in November without checking to see what waterfowl I may find.

5

Winter—
December through February

As winter begins, cold weather settles in, trees are mostly leafless, and the sun sets before 5 p.m., my hibernation instincts threaten to take over. I want to come straight home from work, read a book next to the wood stove, and sleep 12 hours each night. Some years it's even worse, and I really feel like a bear. I crave lots of high-calorie food; bears need to store up fat for the season. Although a bear can grow a heavier coat, I need to add layers of clothing; out of storage come the woolen sweaters and jackets. I spend time in cozy sheltered places, just as a bear seeks its winter den. And sometimes it

Horned larks, snow buntings, and longspurs in cornfield

85

seems as if I could easily curl up on a cold December evening and snooze until February brings longer, warmer days once again.

Much of the natural world rests at this season, too. Many animals pass the coldest times in a dormant state, either in true hibernation, when body systems slow dramatically, or in a similarly quiescent torpor. Bears, bats, chipmunks, woodchucks, turtles, frogs, snakes, and even some butterflies lower their metabolic rates in winter, waiting patiently for winter's silence to be broken by spring's songs. Few plants show much winter growth, most trees leafless and quiet, most herbs withered and brown, but ready to start again in spring with new shoots and leaves. Winter is a resting season, a time of stillness and quiet in the natural world.

Or so it seems. Take a closer look, and it quickly becomes obvious that not everything sleeps through the winter. There's much activity for a naturalist to explore, and I always find myself fighting that hibernation instinct, brewing the morning's coffee a little stronger, bundling up snugly, and heading out on field walks regularly. There are many great places in our region I try to visit each winter, many activities that are part of my winter routine.

Early winter is the season for the National Audubon Society's Christmas Bird Count, a traditional activity all across the country. The data gathered from these counts are significant, showing which of our winter species are declining and which are holding steady or increasing. I enjoy participating in the counts, in part, because I like to make my small contribution to ornithology. In many ways, however, the counts are just as important because they lure me out into the field. Even though it's often hard to get started on a cold December morning, once outside I am never disappointed and am usually rewarded with some memorable sighting. Once I watched a peregrine falcon chase an American kestrel across the Chesapeake Bay. Another year I found seven species of woodpeckers on the Mason Neck National Wildlife Refuge in the first two hours of the day. I'm always pleased if I happen to find a winter wren, hermit thrush, American tree sparrow, or another of our less common winter residents. I cherish many Christmas Bird Count scenes, from a great horned owl silhouetted against the orange sunrise sky to endless flocks of cedar waxwings filling the hedgerows of Hughes Hollow. Most years I participate in four or five different counts; our region has dozens, and newcomers are always welcome. Inexperienced birders are teamed with skilled veterans.

Wetland refuges along the Coastal Plain remain good for birding through-

Short-eared owl over marsh

out winter, as many ducks and geese typically spend the entire winter here. Tundra swans are abundant in the Chesapeake Bay region during winter. Northern harriers and rough-legged hawks may be seen hunting over Eastern Shore marshes; toward dusk short-eared owls often appear. Some of my fondest winter birding memories combine the austere scenery of a winter marsh with sunset colors and the purposeful flights of these graceful, crepuscular owls.

Bald eagles may be observed in many places during winter. Populations of these impressive birds have rebounded significantly since DDT was banned in 1972. Blackwater National Wildlife Refuge, near Cambridge, Maryland, is justly famous for its eagle concentrations, with one or two golden eagles often present along with their white-headed relatives. Closer to the city, watch for bald eagles along the Potomac River, especially south of Alexandria. Several dozen winter around the Mason Neck peninsula, east of Lorton, Virginia. An increasing number of bald eagles are nesting in our region; the nesting cycle begins in winter. Eagle nests are large masses of sticks placed high in a sturdy tree, often adjacent to water, and typically used for many consecutive years. Observe any nest you may find from a discreet distance because nesting eagles are sensitive to human disturbance. Well-watched pairs have nested on the grounds of Mount Vernon, quite visible from the Riverside picnic area along the Mount Vernon Parkway, and on Conn Island, in the Potomac just above Great Falls.

Experienced birders flock in winter to places where gulls congregate, as a number of rarities usually occur in our region. Favorite places for gull watching include a variety of locations along our rivers, bays, and the Atlantic Ocean, plus Georgetown Reservoir, Conowingo Dam, and those great birding hot spots, sewage treatment plants and landfills. A black-headed gull was regularly seen a few years ago at a fast food parking lot in Ocean City, and a Ross' gull made headlines when it visited Baltimore's Back River Sewage Treatment Plant. Ring-billed, herring, and great black-backed gulls are our most common species, but with luck and persistence you might find Iceland, glaucous, Thayer's, little, or lesser black-backed gulls. Laughing gulls, common in our region during summer, are rarely seen in winter. A yellow-legged gull has been seen in our area in recent winters, usually at the Georgetown Reservoir or at one of the landfills.

Once I am coaxed outside in winter for a Christmas Bird Count or other field trip, I find much to observe in addition to the birds. This is a tough season for most of our plants and animals, and many are either dormant or scrambling to sustain themselves. It's a great winter activity to observe the adaptations or behaviors that help each species survive. A few plants and animals begin the new year's reproductive cycle during winter—most of us watch for these hints of spring with fervent anticipation.

Visit a moist forest in January for some of spring's earliest signs. Skunk cabbage's speckled maroon-brown flowers and new leaves, which begin as odd-looking green cones, usually emerge this month. The heat produced as a by-product of this plant's metabolism can melt through a layer of snow. Many trees have buds that gradually enlarge all winter long, ready to burst open with flowers and new leaves in the early spring. By the time skunk cabbage blooms, the red maple trees' swollen buds add red tones to the landscape and spicebushes' yellow-green spherical buds are conspicuous at eye level in the understory. In drier forests look for the long, pointed buds of the beech tree, which is more easily identified by its smooth gray bark. Oak trees cluster buds toward the branch tips, and hickories have very large terminal buds. With practice and a little patience, all trees can be identified by characteristics of the buds and twigs.

Gray squirrels in our area molt during autumn into a winter pelage, a fur coat that is pale gray whereas the summer coat is rich brown. Look for bright white tufts of fur behind a squirrel's ears; the summer coat lacks this feature. Our gray squirrels breed twice a year. In winter they dwell in tree cavities, not in the leafy nests called *dreys*, which are often used in summer. Gray squirrels typically court during late December and early January with an animated set of calls and chases. Watch interacting squirrels to see if you can differentiate between a courtship chase and a confrontation between two males. Confrontations are usually completed more quickly, with the submissive male retreating. Courtship chases feature relentless pursuit of a female by one or more males. Try to be sympathetic if squirrels raid your bird feeders; food is scarce most winters, and squirrels just don't understand that your sunflower seeds aren't meant for them.

Many other mammals are well-adapted for the moderate mid-Atlantic winter, their insulating fur holding the internal heat generated by digestion and respiration. As long as there is food, mammals will stay warm. Bats, however, because they feed on flying insects, either hibernate or migrate once late fall's chill eliminates this resource. A few other mammals in our region also hibernate, but most remain active through the winter. Gray squirrels and white-tailed deer, two of our most readily observed mammals, feed furiously through fall, storing energy for the winter either as body fat or food caches. They must get plenty of food, as both species court and mate during winter. Because the mating cycle places extra energy demands on an animal, most species court at times of good nutrition and high food supply.

Frogs, salamanders, turtles, lizards, and snakes pass the coldest times by reducing activity, finding shelter in a warmer spot (underground or in the mud of a pond bottom, for example), and entering a dormant state. Those reptiles and amphibians in aquatic environments stay active later in fall and return to activity earlier in the spring, but few are found during winter's coldest times. Mid-Atlantic herpetologists find little to study in the field during winter's coldest days.

Many mid-Atlantic locales make delightful destinations for winter excursions. The C & O Canal remains a favorite place for me, year after year, in every season. Winter brings less recreational use to the canal's towpath, a plus for nature study. The great trees along the Potomac shores look noble without their leaves; sycamores, walnuts, and elms have shapes that I particularly enjoy. These old trees harbor woodpeckers all year, but it's much easier to spot these birds in the open canopy of winter. Pileated woodpeckers—crow-sized, red-crested, and noisy—are especially common here; the Seneca

Christmas Bird Count often tallies more of these wonderful birds than any other count in the country. It's easy to find five other woodpecker species at the canal in winter—these include our winter visitor, the yellow-bellied sapsucker, as well as our year-round resident flickers, and downy, hairy, and red-bellied woodpeckers. Our seventh species, the red-headed woodpecker, is rarely found.

The extensive forests lining the Piedmont Potomac River, accessed by the C & O Canal in Maryland and a variety of parks in Virginia, support many other birds in winter. Eastern bluebirds flock during this season and forage through the woods, as do yellow-rumped (myrtle) warblers and cedar waxwings. Golden-crowned kinglets and brown creepers can sometimes be plentiful, each giving weak, high-pitched trills from the trees. It's easy to find Carolina chickadees, tufted titmice, white-breasted nuthatches, northern cardinals, and white-throated sparrows. A lucky day could bring winter wrens, hermit thrushes, tree sparrows, and red-breasted nuthatches into view. Some years we experience a phenomenon known as a winter finch invasion: an influx of finches that usually stay to our north all winter. Biologists believe that these finch invasions occur during years when the conifer trees of the northern forests produce poor seed crops. During these years winter may bring many evening grosbeaks, pine siskins, purple finches, or sometimes even a few red crossbills or common redpolls into our region.

The Potomac River corridor is wonderful for nature study downstream from Washington, D.C., too. The Mount Vernon Parkway provides several fine access points to the river, which is broad and tidal from here to the bay. The Belle Haven picnic area and adjacent Dyke Marsh are prime spots for waterfowl viewing. The Riverside picnic area, overlooking the mouth of Little Hunting Creek, can be a fine spot for observing tundra swan, Canada goose, lesser scaup, canvasback, ruddy duck, northern pintail, and other ducks.

Mason Neck is just a few miles farther downstream. Most of this peninsula is covered by a young forest that features a variety of oaks, hickories, and, in the understory, extensive stands of holly, a traditional favorite for holiday decoration because it stays green through winter. These trees grow and photosynthesize through the coldest weather, taking advantage of the sunlight that penetrates the leafless forest canopy. Several other green plants occupy the same ecological niche in our deciduous forests, including Christmas fern, spotted wintergreen, and partridgeberry.

In early winter you'll crunch many leaves underfoot when walking Mason

Neck's trails, but a welcome silence rewards most stops. I relish the stillness of the winter forest, living in the noisy suburbs as I do, but I also welcome the sudden crashing sound of a deer charging through the woods. Little sounds that are often buried in summer's cacophony stand out in winter, such as the churring of a gray squirrel, the tumbling of water through a tiny creek, and the creaking of tree limbs as they sway in the breeze. The mixture of calls from a mixed flock of wintering songbirds carries a long way through these woods. Because of the plentiful holly berries and the expanse of Mason Neck's forests, the area's winter songbird population is quite high.

Coastal salt marshes are beautiful in every season, but I like them best in winter. Extensive stands of cordgrass wave in the wind like the sea itself. Once rich and green, these marshes have now turned brown, a color that perfectly complements the textured gray sky that typifies cold winter days. Dabbling ducks burst into flight from pools within the marsh, while flocks of geese and cormorants pass overhead. Saltmarsh sharp-tailed and seaside sparrows pop into sight once in a while, as clapper rails slither through the vegetation. Northern harriers drift slowly over the vast open wetland, replaced by short-eared owls late in the day. Similar scenes may be enjoyed at the Chesapeake Bay's brackish marshes, including Elliott Island and the Blackwater National Wildlife Refuge.

The mountains warrant a winter visit, too. If we get a long cold snap I try to visit Shenandoah National Park, Cunningham Falls State Park, or another spot where big waterfalls may be found. I find few scenes in this world as beautiful as a frozen waterfall on a sunny winter day. Shenandoah's mountains are often dusted with snow, and the park's fire roads are ideal for cross-country skiing. Sometimes there's too little snow for skiing, but even one inch opens a window to the park's wildlife, with tracks of deer, rabbits, foxes, bobcats, mice, voles, turkeys, or grouse waiting to be followed.

You're not likely to find a bear's tracks in the midwinter snow. Although bears have become fairly common again on the Blue Ridge, these burly animals spend most of this season snugly curled up in a den. In early spring they'll be out foraging again. I envy elements of the lifestyles of several animals: Arctic terns, which migrate between extreme latitudes of the Northern and Southern hemispheres, maximizing their exposure to daylight; and Olympic marmots, which live in the gorgeous alpine meadows of Washington's Olympic Mountains and hibernate through the long cold winter. By nature, however, I'm more like one of Shenandoah's bears, trying to fatten up in fall and snooze through the coldest part of winter. I try to fight that instinct every year, in order to see all of the wonders those poor bears are missing.

MIDWINTER

There is ambiguity in each of our seasons. Signs of spring can be found throughout winter—skunk cabbage blooms in January as great horned owls complete their courtship and lay eggs. February brings the first maple blossoms and the northbound flights of the earliest migrant waterfowl flocks. A few days of mild weather can come at any time, accompanied by the spring songs of Carolina chickadees and flights of mourning cloak butterflies. But cold spells dominate the season and profoundly influence all the organisms and ecosystems in our region.

The hallmark of midwinter is a spell of exceptionally frigid weather. Cold, dry pockets of arctic air that drop temperatures to well below freezing are brought south a few times each winter by big cold fronts, often given colorful names such as Siberian Express or Alberta Clipper. Some winters bring long stretches of subfreezing weather, but in most years these spells are short-lived and infrequent. Almost every winter brings at least a few such days, when the high temperature stays in the teens or low twenties and the nighttime low drops to 10 degrees Fahrenheit or below.

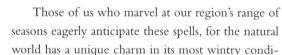

Mourning cloak

Those of us who marvel at our region's range of seasons eagerly anticipate these spells, for the natural world has a unique charm in its most wintry condition. It's a challenge to explore this world comfortably, but today's wonderful insulating clothing really works, and even the underprepared can enjoy a series of short excursions. Whether you stroll for a half-hour at a nearby park or hike all day in the mountains or down the beach, a midwinter exploration can be rewarding.

Snow brings special charms. Animal tracks and trails are subtle and hidden to all but the most careful observers for much of the year, but snow cover makes them obvious. Simply identifying the makers of tracks can be fun, but it's more challenging to try to understand what the animals have been doing. Take to the woods a day or two after a snow and try to read the stories:

Where have animals traveled and bedded down? At what point did an animal switch from a walk to a run? Can you find pairs of animal trails telling the story of pursuit and predation or escape? Will a line of tracks lead to an animal's den? Can you find places where an animal has scent-marked its territory with its droppings or scratchings? A snow-covered natural area is like a detective novel, with lots of clues in its many plot lines.

Winter birding can bring great gifts, with or without snow. Many forest birds congregate in loose flocks at this season. You can walk through the woods for half an hour without hearing a peep, then suddenly find yourself surrounded by a group of little birds. Sometimes you'll hear just a few small chips or squeaks, but some creative noise making can stir the flock into action. Try *spishing* (making hissing noises by pushing air through clenched teeth) when tufted titmice or Carolina chickadees are nearby and you may find a number of birds coming close to investigate. There's likely to be a Carolina wren or two and a downy woodpecker nearby, and sometimes several other species as well. Golden-crowned kinglets, brown creepers, yellow-rumped warblers, white-throated sparrows, red-bellied woodpeckers, white-breasted nuthatches, and northern cardinals are among our most common winter forest birds.

Lakes, rivers, wetlands, and bays are busy spots throughout winter. Many kinds of ducks, geese, and swans winter in our region, along with loons, grebes, coots, gulls, and great blue herons. I'm always amazed to see birds sitting calmly on frigid waters—why don't they freeze? Their two keys to warmth are fuel and insulation. The fuel is food, as heat is a by-product of digestion. Fat layers and downy feathers (groomed carefully with water-shedding oils) provide insulation. Healthy animals that have adequate food in winter generate all the heat they need internally. In other words, those ducks on the lake are biological blast furnaces wearing down jackets.

Winter's coldest times differentiate our temperate ecosystems from tropical ones. Summers bring us hot, humid weather, always well above the freezing point. In summer our forests can resemble those of the moist tropics, except that we have many fewer species of plants and animals because tropical ecosystems are not limited by freezing weather at certain times of year. Relatively few organisms have adaptations that enable them to survive long spells of severe cold. Most animals suffer when their sources of drinking water freeze; this is true for Carolina wrens, for example. These active little nonmigratory birds are near the northern edge of their range here, and unusually cold winters bring severe mortality to their population. Lack of open water for drinking seems to have a major negative effect on this species.

Most plants and animals enter a dormant state during winter, but even in dormancy there is a lot of mortality. Many frogs, snakes, woodchucks, and insects quietly expire during winter's coldest times. Large numbers of birds, along with some bats and insects, migrate away for the winter, heading to warmer areas where food continues to be available. Migration has many hazards, though, from its incredible energy demands to the threats of predation all along the route. A lot of our migrants who travel south in fall won't come back in spring.

Many resident animals are increasingly vulnerable to predation in winter as they weaken from cold, inadequate food, lack of open water, and a decline in vegetative cover. The annual cycle of great horned owls reflects this. Birds need the greatest amount of food when they are feeding young at the nest. Consequently, their breeding cycles have evolved so that hungry nestlings are being nurtured at the time of greatest food availability. For great horned owls, which feed primarily on small- and medium-sized mammals, this time is late winter, when many animals with marginal resources for winter survival are weakest.

Death is everywhere in our winter environment. Most creatures of the temperate zone go through annual population cycles, with expanding numbers in summer and declining ones in winter. To understand ecology it is critical to appreciate the role of death in an ecosystem. To convince others to accept the role of death in nature is also one of the most challenging aspects of environmental education. Faced with the death of an animal, many people want to intervene. Someone finds a baby bird on the ground in spring and wants to nurse it to fledging. "Let it go," I'm tempted to say. "It's nature's way for some reproduction to fail." Another person will see predation in action, a snake raiding a nest or a hawk taking birds from a feeder, and call in alarm. I'm asked, "How can the prey be protected?" Again my inclination is to say, "It's nature's way." There is a vital relationship between predators and prey in nature.

When predators are removed from an ecosystem, survivability within the population of prey animals increases. Weak animals that would normally have been the first ones taken by predators are able to reproduce, sometimes passing weak characteristics or behaviors to the next generation. Eventually the prey animals may overpopulate, feeding intensively on their food resources until these are depleted, followed next by catastrophic population crashes. Before the decline, however, peripheral environmental damage is likely to have occurred, affecting the populations of many other plants and animals. The numbers of animals in any given environment are always fluctuating, but the levels remain much steadier in ecosystems where predators are still present. To protect animals it is best to protect habitats and all their component organisms, places where the dramas of predation and a complex ecosystem can continue as they have for millennia.

But when I hear caring and true concern in the voice of another, I stop to reconsider my advice. Sure, it's important to understand ecology and to accept the role of predation and mortality. But do I want to encourage people to ignore their compassion, their reverence for life? I don't think so. All of us who cherish the natural processes on this Earth need to develop a balance between our cognitive acceptance of death and our emotional compassion for the lives of our fellow inhabitants on the planet. I find myself responding to questions from the rescuers of injured or orphaned young animals with a mixed message. I try to give the ecological message and temper expectations ("Sadly, even under the best care this animal isn't likely to make it," I'll say), but I encourage their compassion and direct them to experienced rehabilitators or appropriate reference books. Caring is the first step, understanding the second, and empowerment the third in the process of environmental education.

Winter mortality isn't good or bad, it is natural and inevitable. It's part of the ecological drama that began with the first living creatures on Earth. To fully appreciate midwinter we need to understand its role in the great cycle. The stage is being set for the riotous explosion of life that comes every spring. Whenever winter gets me down I'll head down to wet woods where skunk cabbage grows and greet those early flowers. They're like mystical crystal balls, providing glimpses of the future—or perhaps they are more like a pulse, a place where one can feel the vital rhythms of this living planet.

LATE WINTER

Most of us like to celebrate the new year, looking forward to the next cycle of seasons in our lives. From a naturalist's perspective, I'm never sure when the new year begins. I figure it's not the calendar's arbitrary day, but rather some late winter day when the world is clearly moving toward spring. In some ways, the midpoint between the winter solstice and the spring equinox, which falls around February 7, seems like the best choice, for by this time the expanding day length is nurturing spring activity in many of our plants and animals.

Ah, those longer days! Late winter can bring us winter's coldest weather or deepest snows, but whatever the weather, we know that spring is coming because the sunrise comes earlier each day, the sunset later. I'm still inclined to get out of town and head to tropical destinations, or to stay home and read, but longer days help me overcome my hibernation instincts. Whether the late winter is cold or mild, the increased light coaxes me outside more frequently than in early winter.

Spring is just around the corner by the end of February, and the progress toward this renewing season can dominate my natural wanderings during late winter. Buds on red maple trees swell throughout the month, their rich, burgundy color standing out clearly against the gray background of the winter forest. A warm spell near the month's end can bring quite a few maple flowers into bloom. Bud expansion is evident in other species of maple, as well as in spicebush, elm, and several other woody plants. Visit a weedy, south-facing lawn after a few sunny days in February when the temperature rises above 50 degrees Fahrenheit and you're likely to find the tiny flowers of a few common herbs, such as common chickweed, purple dead-nettle, henbit, or speedwell. Even dandelions sometimes bloom in February. The year's first flying insects may also be seen on such a day. A few of our butterflies winter as adults, though they spend most of their time in a dormant state. Be delighted but not surprised, if, on a warm and sunny February day, you spot a mourning cloak, eastern comma, or question mark darting through an open forest patch.

Wood frogs are the first of our frogs to court and breed, often becoming active during warm spells in February. Their rough squawking calls can be heard during afternoon and evening hours. Watch for them around quiet pools of water in forested areas. It seems like there are one or two days in the late winter or early spring when all the wood frogs mate. If you visit a pond on one these days you may see dozens of paired frogs floating and drifting around, with four little bulging eyes poking out of the water. Wood frogs will breed in permanent ponds of varying sizes and in temporary pools. Once at Hughes Hollow I found wood frogs breeding in the small puddles formed by tractor treads in the low, saturated ground.

A warm spell in February can also bring spring peepers out of dormancy. When these abundant little treefrogs have emerged they head to ponds and marshes where they'll call all night. Their shrill, upslurred whistles are one of the surest signs of spring.

WINTER BIRDING AT THE BEACH

Most people love warm summer days at the beach. I go to the Atlantic coast throughout the year, but I especially love it when the weather is cold. There's something I really like about dressing up in all my warm clothes and heading onto the beach, grimacing as cold gusts blast my face with windblown sands. Often the winter seas are churning, pounding the beach with powerful breakers. The world rarely shows its powers as emphatically as on a stormy winter day at the beach. Although there are interesting birds to see in all seasons on the coast, I'm especially fond of winter birding here. I plop down on the sand, hopefully at a spot where the dune offers a bit of a windbreak, and gaze intently out to sea. I hope to find common loons, red-throated loons, buffleheads, red-breasted mergansers, oldsquaw, any of the three species of scoter, and horned grebes out bobbing on the ocean. I scan back and forth along the horizon in search of northern gannets; some days the ocean seems empty, but on others I see hundreds of gannets flying by as if in a parade.

The dunes behind the beach always warrant scrutiny. Snow buntings, one of our most beautiful sparrows, often move along the sparse dunes in flocks

Surf scoters

that can range from two to two hundred. A big, pale race of savannah sparrows (known as the Ipswich sparrow) enjoys our winter dune habitat. At many coastal areas wax myrtle shrubs form dense thickets inland from the dunes. Here, hardy songbirds spend the winter, including huge numbers of yellow-rumped (myrtle) warblers. Some birders grow tired of this abundant little songbird; I've heard people sigh many times, "Just another myrtle." I'm always amazed at the tenacity of this little bird, a member of a primarily insect-eating clan. Most warblers retreat to the tropics where insects fly year-round, and their migratory feat is truly extraordinary. I'm equally impressed with the myrtle, however, which has adapted to feed on fruits during the winter and is able to thrive under austere conditions.

Raptors frequent the coastline in great numbers and diversity during migration, and a good selection can usually be found throughout the winter. Snowy owls venture south into our region some winters, and coastal areas, especially open dune regions, are among their favorite locales. Red-tailed hawks, rough-legged hawks, northern harriers, sharp-shinned hawks, peregrine falcons, and bald eagles may all be observed throughout the winter.

My favorite places to explore the winter beach are at Assateague Island and at Cape Henlopen. In addition to a lovely wild beach, Assateague's southern end, managed as the Chincoteague National Wildlife Refuge, features extensive wetlands where waterfowl watching is outstanding from November through February. Few scenes match the grandeur of a winter sunset over the refuge's freshwater marsh with thousands of snow geese in the sky. Up by Cape Henlopen, a side trip north to Broadkill Marsh offers a show with even greater numbers of geese. I automatically end every winter trip to Cape Henlopen with a sunset visit to Broadkill.

WINTER NIGHTS

Snow cover and a full moon are an intoxicating combination. Moonlight pours through the open forest canopy and reflects from the snow, brightening the scene. The winter woods are quiet without the raucous calls of frogs and crickets so common for much of the year, and snow seems to muffle the few existing sounds. With or without snow a winter night walk may give you a chance to hear great horned owls calling *hoo-huh-hooo, hooo, hooo.* They court and begin to nest in December and January.

Stars are best studied in winter thanks to short day length, frequent clear nights, and an abundance of bright stars. Watch for the constellation Orion in the southern sky and the nearby star Sirius, part of the constellation Canis Major (the Big Dog). Sirius is the brightest star that can be seen from Earth, though the planets Venus, Jupiter, and Mars are sometimes brighter. Sky watching has many rewards, some unpredictable. Shooting stars are meteors, bits of cosmic debris burning up in our atmosphere. Occasionally a big chunk burns spectacularly; I'll never forget the time I watched with awe as a brilliant green light streaked through the sky, breaking into a dozen distinct glows before disappearing behind the wooded horizon. Satellites can often be seen as faint starlike lights that move steadily across the sky. All of these celestial objects are best observed on moonless nights.

WATCHING THE BIRD FEEDERS

I made a bird feeder years ago as a Boy Scout project. I banged together a few pieces of scrap wood and bought some canary seed from the neighborhood grocery store. A few house sparrows came to my feeder, which was all I needed to get this requirement signed off on my card. Bird feeders were considered good projects for kids in those days, and a few serious bird watchers would set out a feeder or two for the winter, but feeders were not so common as to influence bird populations and behaviors.

Now, bird feeding is big business; millions of people have feeders, and many million tons of bird seed are sold each year. Feeders help increase the winter survivability of many native year-round resident birds, including Carolina chickadee, tufted titmouse, northern cardinal, and mourning dove. Our winter residents also benefit from this abundant food resource, including birds such as dark-eyed junco, white-throated sparrow, and pine siskin. On the negative side, feeders are thought to have nurtured population increases for several nonnative birds, particularly Eurasian starling, house sparrow, and house finch. These birds sometimes compete for resources with native species. Good or bad, I believe the bird-feeding craze is here to stay. During chilly winter spells and the season's harsh storms, watching bird feeders can be a fine nature study activity.

A good feeding station features several types of food. Many songbirds prefer sunflower seeds or mixtures that include sunflower and millet. American goldfinches and pine siskins prefer thistle seed. Suet and peanuts attract woodpeckers and wrens. Because many natural water sources freeze during cold weather, a heated birdbath is a valuable adjunct to a feeding station. If you don't want to bother with a heater, you can set fresh water out near the feeders once or twice every day during freezing spells.

You are likely to attract more birds if you place your feeders near shrubs and trees that provide hiding places and shelter. After a while you're likely to have a group of birds whose winter territory is centered around your feeders. At this stage it's crucial to keep food in the feeders through the winter, for these birds are now dependent on this food source. Without it, many are likely to perish. If you plan a vacation, find a friend who can fill the feeders while you're away.

Most of the birds that visit feeders are fairly common, but once in a while an unusual bird may appear. Purple finch, pine siskin, and red-breasted nuthatch are species I hope to find at my feeders each winter. Every few years evening grosbeaks return to our region for the winter; if they are in the area, these big yellow and black birds are likely to find any feeder with sunflower seeds. Common redpolls will also come into our region every once in a while during winter, and they usually join the goldfinches at thistle feeders.

Sometimes activity suddenly drops around bird feeders that had previously attracted many birds. When this happens, it's likely that a sharp-shinned hawk or a Cooper's hawk has discovered the feeder. Both are winter residents in the mid-Atlantic, and both feed on songbirds, typically taking the weakest songbirds in a flock. Once a few birds have been caught by a hawk, the others may move to another nearby feeding station. A few days later, the hawk may find the flock again, and then a few days later the flock may move back to the first feeders. When hawks arrive, a backyard bird feeding station turns into a dynamic ecosystem.

I put out my bird feeders during October and take them down in April. Some advocate keeping feeders active all summer, though I suspect that the most vehement advocates of this practice are those who sell the seed. Certainly birds will continue to visit feeders throughout the summer, but I fear that when we feed during the nesting season, we sometimes greatly increase the reproductive success of nonnative birds such as the house sparrow, which then compete with native species for nesting sites and for food. The aggressive house sparrow will actually kill young eastern bluebirds to take over a nesting box or cavity. For this reason, I choose not to feed during the nesting season.

Places to Explore
Six Journeys through the Mid-Atlantic

The mid-Atlantic region has countless areas where nature can be observed and explored. In this section I follow six paths through the region, stopping here and there to hint at the wonders you might discover with a visit. I want to leave you with plenty of discoveries to make for yourself! My selection is rather random, based primarily on my own experiences; consequently, there are just as many interesting natural areas that I've omitted as ones I've included. With this section, I hope I give you ideas of places to go and explore natural history and to help you understand some of the relationships among many of the interesting natural areas of our region.

We will travel first to West Virginia by following Highway 55, known as the Highland Trace, to a rich assortment of natural areas in the highlands of this state. Next we will visit the Potomac River, whose path touches mountains and lowlands, wild places and the city. After that, we will traipse over the Ridge and Valley region, tracing the path of the Appalachian Trail through the mid-Atlantic and visiting other areas a bit farther west. Our fourth journey will be to the Chesapeake Bay, from its beginnings as the Susquehanna River to its mouth at Hampton Roads. Wetlands and other natural areas close to the Chesapeake are also included here. From the Chesapeake, we will wander down to the Atlantic coast, looking at areas from New Jersey to the southern edge of Virginia, as well as Delaware Bay. We will stay close to home in our last trip and examine many of the parks, refuges, and other natural areas close to the cities of Washington, D.C. and Baltimore.

It's my hope that these passages tempt you out into the field to explore. A number of outstanding guidebooks already exist, many of them listed in the reference list; most of these books provide precise directions to places I discuss here. Whether you choose to explore on foot, from a car, or from your armchair, I wish you a pleasant journey!

6

West Virginia Highlands— Nature down the Highland Trace

I have sketchy recollections of the many times our family traveled to West Virginia during my youth. We always headed to Richwood, Webster Springs, and thereabouts, where my grandfather was born and raised. I didn't always enjoy the visits to distant relatives' homes or long stops at rural cemeteries where the tombstones were dominated by family names. I certainly didn't like the "hillbilly music" that was played at all special events, though today I love that music, which is now trendy and called "bluegrass."

What I always did enjoy was the journey. We traveled on winding roads that pierced dark forests in the West Virginia Highlands, through the pastoral beauty of the upper Potomac River, and past the daunting towers of Seneca

Seneca Rocks

Rocks. I remember strolls along rocky mountain streams, where I'd hope to glimpse a beaver or a big trout, and the expansive scenic views from the ridgetop roadsides. What wonders must have been hiding back in those woods, up on the hills, and down in the hollows.

In college I latched onto a group of hikers determined to find out about those hidden wonders. We headed to West Virginia several times a month, hiking trails, canoeing rivers, exploring caves, and climbing cliffs. Soon I started to notice the wonderful wildflower displays and diverse wildlife communities of these mountains. I've been returning frequently ever since.

The West Virginia Highlands combine diverse natural communities with beautiful, quiet surroundings. In the summer months, when much of our region is a steamy inferno, these highlands remain cool and comfortable. Most areas are never crowded, inexpensive meals and lodging are easy to find, and the 3- to 4-hour drive from the city is scenic and generally free of traffic. What else could a naturalist want?

My West Virginia explorations usually begin with a drive around the Beltway to I-66, followed west to its end near Strasburg, Virginia. Except during the afternoon rush hour (avoid at all cost!), this is a quick journey. Once beyond the suburban outpost of Manassas, the highway winds through scenic Thoroughfare Gap, a strategic place in the Civil War battles of Bull Run. Here, above the banks of Broad Run, a regionally rare butterfly called the gold-banded skipper can sometimes be found. I always like to see the lovely old stone mill that lies just north of the road, tucked into a small patch of forest where yellow warblers and eastern phoebes nest.

From Strasburg my journey continues on the two-lane Virginia Route 55, which crosses the state line and becomes West Virginia Route 55 on the top of Great North Mountain, about a dozen miles west of Strasburg. A ridgetop hiking trail leads south from the boundary through the dry forest that is typical of the narrow mountaintops of the Ridge and Valley. A quiet little U.S. Forest Service campground called Hawk lies to the north. Nearby is the quiet Capon Springs Resort, a throwback to the early decades of the twentieth century. The waters of Capon Spring have long been noted for their healing powers; stop by the resort for a drink or, in season, strawberry pie.

As Route 55 crosses into West Virginia, it picks up the label of Highland Trace, thanks, no doubt, to more poetic members of the West Virginia Highway Department. After a long descent Route 55 reaches the town of Wardensville, where I've enjoyed good country victuals at several small restaurants. Wardensville's six-block Main Street is classic Americana, pretty as a postcard.

County Road 55/1 is a little looping side road to the north of the Highland Trace just east of Wardensville; it runs along the base of a small shale barren. The underlying geology of eastern West Virginia is primarily sedimentary rock of Paleozoic age, some 200–400 million years old. Limestone, sandstone, coal, and shale rocks are often exposed at the surface in mountainous country. Where shale occurs on a south-facing slope, an unusual plant community called shale barren occurs; this habitat is scattered throughout the Ridge and Valley province. The Wardensville barren is protected as a Nature Conservancy preserve, and its plants should not be disturbed. Fortunately, the barren and its plants are easily observed from this gravel road; plants include uncommon ones such as the Kates Mountain clover, shale onion, and everlasting groundsel. Watch out for the big copperhead that hides under the boulder near the southwestern base of the barren.

Kates Mountain clover

The Highland Trace next climbs over the northern end of Buck Mountain, part of the George Washington National Forest, and descends to the Lost River. What is lost? The river itself was lost by early settlers, who found, no doubt to their amazement, that this river disappears into the mountain. People later realized that the river emerged on the other side, where it is known as the Cacapon River, but the upstream side still retains the name Lost. There's a small picnic area next to the river, from which you may walk downstream a short distance to the spot where the river sinks into the ground. In a dry summer there's little water at this spot.

Lost River State Park is near the river's headwaters in an unusually moist section of Ridge and Valley forest. From the Highland Trace, take State Route 239 south from Baker to reach the park. Follow some of the forest hiking trails here to search for a good variety of birds, wildflowers, salamanders, and trees. Cozy cabins may be rented in the park. Route 12, a quiet gravel road leading north from the park toward Moorefield, crosses limestone

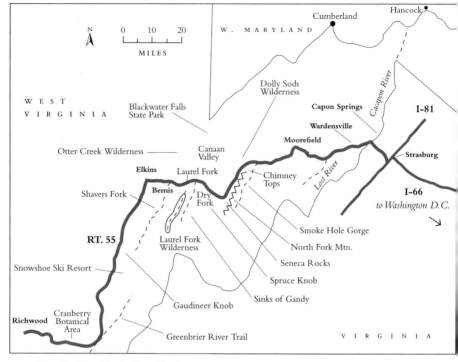

Map 2.   West Virginia's Highland Trace

ridges where the abundant wildflower display of late spring features uncommon species such as shooting-star and birdsfoot violet.

Beyond the valley of the Lost River, Route 55 crosses several high ridges. Don't try to hurry along this stretch of road, and believe the orange caution signs that warn of safe speeds for upcoming turns, in places as low as 15 miles per hour. Planning has begun to extend Interstate 66 through this section of the mountains. Most of the locals oppose the new road, believing it will change the character of the region dramatically, in both natural and cultural ways. Count me among those who would prefer no change.

Moorefield is the next major settlement along the Highland Trace, reached about 30 miles beyond Wardensville. Its welcome signs proclaim the little city as the "poultry capital of West Virginia." Try the chicken entrees at any local eatery; you're not likely to overpay. Moorefield sits in the broad valley of the South Branch of the Potomac River, surrounded by big farms and little towns with names like Fisher and Rig. The surrounding mountains here and to the west, which include some of the mid-Atlantic's highest peaks, are known as the Potomac Highlands. Many outstanding natural areas are found in this region; on most trips to West Virginia, I don't stop until I've made it past Moorefield.

It's a quick 12 miles southwest through the valley from Moorefield to Petersburg along the Highland Trace. A few miles west of Petersburg lies the office of the Potomac District of the Monongahela National Forest. Helpful information is available here during business hours, and free brochures may be picked up outside the front door at any time of day. A few miles farther along, on the right, is Smoke Hole Caverns, where the intricacies of limestone caves may be comfortably observed on guided tours. Even though the Smoke Hole Caverns' business claims to have West Virginia's largest souvenir shop, I've never explored its selection. The many billboards will ensure that you'll notice Smoke Hole Caverns as you continue along the Highland Trace.

The Potomac's South Branch forks near here, with Route 55 following the North Fork. The other fork is regionally known as the Smoke Hole, named for the gorge through which it runs. Quiet gravel roads lead into parts of the Smoke Hole Gorge, accessed from the Highland Trace via County Road 28/11, with interesting plants and wildlife to explore along the roadsides. A secluded U.S. Forest Service campground, picnic area, and trail system lie within the gorge. A loop trail of about 3 miles, which leads past varied forest and field terrain from the Smoke Hole picnic area, is a favorite of mine. The river provides a challenging canoe trip through wild and scenic country. Purple cliffbrake, a dainty little fern, grows on the exposed limestone rocks of the gorge, where the brilliant red blossoms of fire pink also may be found. The northern section of Smoke Hole Gorge has limited access and remains unspoiled. Much of the land in this section is protected as a preserve by The Nature Conservancy. Immediately to the west, North Fork Mountain has a long trail on its summit with sweeping vistas west across the Potomac Valley to Dolly Sods and Spruce Mountain. For two years in the early 1990s, peregrine falcons nested here. Two access trails lead to the northern stretch of North Fork Mountain just south of Route 55, providing relatively easy access to the dramatic cliff-edge vista and the majestic pinnacles at the ridge's northern crest, a high point known as Chimney Tops. Views reach beyond the North Fork to an imposing high and long unbroken ridge called the Allegheny Front.

The Allegheny Front forms the western edge of the Potomac watershed and of the Ridge and Valley geographic province. It is an imposing feature when viewed from the east, a steep-walled ridge rising more than 3,000 feet

(900 meters) above the valley floor. West of the front is the Allegheny Plateau. Here, the landscape is dramatically different with broad ridges and narrow stream valleys; in the Ridge and Valley, the ridges are narrow and the stream valleys are deep and broad. The climate is much cooler and wetter in the Allegheny Plateau, with northern forest types dominant.

Climb the Allegheny Front just northwest of the mouth of Smoke Hole Gorge in Petersburg to reach the Dolly Sods plateau. From the Highland Trace 9 miles south of Petersburg, take County Road 4 to either U. S. Forest Service Road 19 or 75; they connect up on the mountain. Dolly Sods is a high, rolling landscape incised by Red Creek and its tributaries. The upper reaches of this plateau are over 4,000 feet (1,200 meters) in elevation, covered with a mixture of red spruce and northern hardwood forest, heath barrens, and bogs. More than 10,000 acres (4,000 hectares) of Dolly Sods constitute a federal wilderness area laced with intriguing trails, but the surrounding areas are equally interesting for nature study. Blueberries are abundant, ripening in late July and early August most years. Beavers have elaborate dam systems on most of the creeks. Memorable finds of summer weekends at Dolly Sods have included black bears, Appalachian seal salamanders, pink-edged sulphur butterflies, golden chanterelle mushrooms, a saw-whet owl calling through the night, and sightings of more than a dozen species of nesting warblers. Trails through the Dolly Sods Wilderness leading up from the southwestern access point of Laneville are justifiably popular, though I always spend more time in the rolling, open highlands farther north, near the rocky summit of Bear Rocks. Visit Bear Rocks in June for a terrific display of blooming rosy wild azaleas; return in late September and watch the migrating hawks streaming by. Head south to the Red Creek campground during fall to see migrant songbirds. The Brooks Bird Club, based in Wheeling, West Virginia, operates a busy songbird banding station just east of the campground. Visitors are welcomed.

If you continue south on Route 55, you will reach Seneca Rocks about 11 miles past the Dolly Sods junction. This sheer quartzite cliff rises a thousand feet straight up, a spectacular sight. Rock climbers come from all over the East Coast to scale some of the more challenging routes. I prefer the easy U.S. Forest Service trail, complete with stairs, that leads along the outcrop's edge. The North Fork of the Potomac River lies at the base of the cliff. Birding is great along the riverside forest, as yellow-throated warblers nest in the sycamore trees and least flycatchers in the maples. Listen for golden-winged warblers where the forest gives way to open field.

Amenities at Seneca Rocks include a large picnic area and a visitor center run by the U.S. Forest Service. This section of the Monongahela Forest is part of the special management area called the Spruce Knob/Seneca Rocks National Recreation Area. Just a few miles to the south of Seneca Rocks is the large, comfortable Seneca Shadows campground, where eastern screech-owl, great horned owl, and whip-poor-will can often be heard calling throughout the night. The little town of Seneca Rocks, formerly called Mouth of Seneca, has food, lodging, gas, camping equipment, and a climbing school. Be sure to stop by Buck Harper's country store. Years ago old Buck told me he would personally guarantee all the cookies he sold in his store; if I didn't like them, bring them back and he'd eat them all!

Spruce Knob, the highest mountain in West Virginia, isn't far from Seneca Rocks. Bear left at Seneca Rocks onto combined U.S. Route 33 east and State Route 28 south and watch for the signs to the turnoff on the right, a few miles past Riverton. A good gravel road winds up to within a stone's throw of the mountain's 4,863-foot (1,482-meter) summit. It's always cool and breezy here, even on summer's hottest day. Most visitors come for the views. To the west is the densely forested section of the Allegheny Plateau that forms the headwaters of the Cheat River and its many forks. The long, high ridge of Spruce Mountain extends both north and south, undulating down to low gaps and back up to other summits. The view east might be most spectacular, though the best viewpoint in this direction is a couple of miles east of the summit, along the roadside. The Potomac Valley lies far below.

Follow the short trail from the end of the road to the observation tower on Spruce Knob's summit. This trail, which continues beyond the tower and then loops back to the parking lot, leads through mixed spruce and northern hardwood forest, where yellow-rumped and magnolia warblers nest. Wild bleeding hearts bloom for much of the summer here. The Spruce Mountain trail leads north from the parking area into wild and beautiful forests and meadows. Other trails lead through scraggly ridgetop forests, where blueberries abound in the understory. Backroads leading west from Spruce Knob cross through a region of open pastures underlain with limestone, an area of extraordinary pastoral beauty. Watch for flocks of bobolinks in these meadows, and see if you can find the Sinks of Gandy, a famous spot where Gandy Creek flows under the mountain through a limestone cave. The little road that winds from Spruce Knob past the Sinks of Gandy eventually leads to the Laurel Fork campground.

Bear right at Seneca Rocks to stay on the Highland Trace. The road climbs 3,000 feet to the summit of Allegheny Mountain, leaving the Potomac River watershed behind and entering the realm of the Cheat. This river, which eventually drains into the Monongahela (and thence to the Ohio and the Mississippi), has four major forks flowing south to north. Traveling west from Seneca Rocks the Dry Fork (with Gandy Creek as one of its tributaries) is first crossed, followed by the Laurel Fork, Glady Fork, and finally Shavers Fork. Each branch is lined with moisture-loving northern forests, thanks to the cool moist climate of the Allegheny Plateau. The Cheat River watershed is exceptionally wet, receiving nearly twice as much annual precipitation as areas such as Petersburg along the Potomac. These branches of the Cheat and the surrounding mountains form a region called the High Cheat.

THE HIGH CHEAT

The High Cheat is managed under multiple-use guidelines as part of the Monongahela National Forest. The dominant forest type is mixed northern hardwood, common trees being sugar maple, black birch, yellow birch, American beech, and black cherry. Conifers such as red spruce, balsam fir, and eastern hemlock are scattered throughout, especially in the cooler places along mountaintops, next to mountain bogs, and in shady valley corners. Little commercial logging of the forest has occurred in recent years, though this could change as the forest stands mature. Natural gas wells in this region have been significantly developed, though the wells and their connecting pipelines don't intrude too much on the forest serenity. An extensive series of U.S. Forest Service gravel roads and foot trails provides access to many ridges and valleys throughout the High Cheat. Signs aren't frequent or always maintained, so take a map when you visit.

Shavers Fork is the largest branch of the Cheat in this region, and its valley holds many charms. There's a great swimming hole just upstream from the town of Bemis, and a seldom-used rail line leads south from here through much of the roadless valley of Shavers Fork south to the little settlement of Cheat Bridge. It's fun to hike along the old tracks, though you can also access the river from trails that connect with the Shavers Mountain Trail, which, in turn, can be accessed from several places along the low-traffic, gravel Middle Mountain Road, which heads south from the Highland Trace at the little town of Wymer. The High Falls of the Cheat, located on

Shavers Fork, is a scenic waterfall surrounded by peaceful forest. Reach the High Falls by following the tracks upstream from Bemis or taking the trail that leads down from the Shavers Mountain Trail.

Federally designated wilderness areas in the Cheat watershed include Laurel Fork, Dolly Sods (Red Creek flows into the Dry Fork), and Otter Creek, which is a wild and rugged area just north of Route 55 near the town of Alpena. I've never managed to keep my feet dry when hiking along Otter Creek; the beavers are constantly rerouting the stream's flow. I've also never been anywhere with as many nesting black-throated green warblers as along the Shavers ridgetop trails in the Otter Creek Wilderness. The wet meadows at the southern entrance to the wilderness are filled with wildflowers and butterflies on early summer days; watch for the silver-bordered fritillary, which is rare this far south, visiting the meadow flowers here. The biggest snapping turtle I've ever seen was up near the northern entrance to the Otter Creek Wilderness, near the town of Parsons.

I hesitate to mention the Laurel Fork Wilderness, for so few people seem to know about the place and it's always peaceful and quiet. Wilderness designation protects a long strip of land lining the Laurel Fork of the Cheat. Along this gentle mountain stream lie pastoral moist meadows and a forest that fills with flowers in late spring. The U.S. Forest Service's Laurel Fork campground is tucked far back into a big section of the High Cheat that has no paved roads. Not surprisingly the site is small yet rarely filled. It's a fine base for a study of this region, though if you're willing to hike just a few miles with a backpack you can find dozens of idyllic spots to camp in complete isolation. This is where I came for three days one year in May to celebrate the completion of my academic studies. Black-capped chickadees, Blackburnian warblers, magnolia warblers, red-spotted newts, and some of the creek's many beavers were my fellow celebrants, but I didn't see any other humans. It was cool and drizzly—it's often wet here—and the green of the valley matched that of a rain forest. In fact, what I really want to tell everyone about the Laurel Fork is that it's always rainy here, and the trails are always brushy and muddy. You don't want to go here (wink). Access the Laurel Fork from the Middle Mountain Road.

The Blackwater River's watershed lies to the north of all the forks of the Cheat listed previously, yet it also flows into the Cheat. Two beautiful state parks, Blackwater Falls and Canaan Valley, are found at the head of the Blackwater River, which begins in the broad, oval-shaped Canaan (accent on the second syllable) Valley, a vast wetland complex dominated by boreal habitats. Canaan Valley State Park protects the southern portion of this valley and provides recreational facilities including a golf course, ski resort, campground, and lodge. Several easy hiking trails lead through the park, past bogs lined by balsam fir trees where alder flycatchers and hermit thrushes nest. Watch for Blackburnian warblers in the surrounding forests. The Canaan Valley National Wildlife Refuge was created in 1994 to protect some of the valley's extensive wetlands and other habitats. In the first tract of land acquired by the refuge, along Freeland Road, there's a large open meadow that fills with wildflowers in midsummer. Here and in other Canaan Valley meadows, watch for nesting bobolinks and northern butterflies such as Atlantis fritillary.

The northern portion of Canaan Valley is primarily owned by the Allegheny Power Company. For more than 20 years conservationists have been successfully battling a plan to drown much of the valley under a reservoir. Congress has authorized acquisition of most of this land for the Canaan Valley Wildlife Refuge, but at this writing the funding has not been approved for its acquisition. Intrepid paddlers can take canoes up into this area along the Blackwater and Little Blackwater Rivers during high water times, though many detours around beaver dams are required. Rough dirt roads lead into this section of the valley; the power company is currently permitting open access to these roads with restrictions posted on the boundary. Be sure to check the signs for restrictions before entering this part of Canaan Valley and, as in many areas, use particular caution during the autumn hunting season.

A steep-walled gorge has formed where the Blackwater River leaves the high Canaan Valley. Blackwater Falls State Park preserves much of the gorge, including the famous waterfall for which the park is named. This is an excellent location for the endangered Cheat Mountain salamander. Nesting warblers that are common in the park include Canada, Blackburnian, magnolia, and black-throated blue. Be sure to visit the little nature center located by a small lake, especially if you're interested in butterflies. One species, the Harris' checkerspot, is very rare in the mid-Atlantic, but it may be found in early summer in the wet meadows

Cheat Mountain salamander

surrounding the lake. Forest Road 13, scenic but quite rough in places, loops south of the park through the Canaan Mountain section of the Monongahela National Forest. A number of hiking trails wind through rich forests in this region. Reach the Blackwater River region by leaving the Highland Trace at Harman and heading north on State Route 32.

After a series of ups and downs past the high ridges of the Cheat Mountain region, the Highland Trace drops down to Elkins, a medium-sized town in the rolling, lower country west of the Allegheny Plateau. Motels, restaurants, and stores can be found here, along with Davis and Elkins College and the headquarters of the Monongahela National Forest. The Augusta Festival celebrates the life and crafts of mountain people each summer with classes and other special events held on the Davis and Elkins campus.

From Elkins the Highland Trace piggybacks with U.S. Routes 250 and 219, which head south along the western edge of the High Cheat region. Side roads head east from the Highland Trace up to Cheat Mountain, where rhododendron displays in June are a floral highlight. At Huttonsville the Highland Trace continues south, but we'll detour east along Route 250, which leads southeast along the southern edge of the High Cheat. After climbing over Cheat Mountain and then descending to cross the Shavers Fork at Cheat Bridge, the road climbs steeply along the south flank of Gaudineer Knob. To the left is a big, but well-hidden, wetland called the Blister Swamp. There are a few pullouts here from which you can venture back into this soggy mix of forest and glade, which gets its name from the abundant balsam fir trees (known locally as blister pines) that grow here. Don't expect to find a formal trail, and be ready for wet feet.

A side road leads north from the next ridgetop, which divides the north-flowing Cheat River drainage from the south-flowing Greenbrier, to the lofty summit of Gaudineer Knob, at an elevation of 4,445 feet (1,355 meters). A dense stand of young red spruce covers the mountaintop, and the fire tower I remember from my first visit has been removed, so there aren't many sweeping vistas from this mountain. There are, however, endangered Cheat Mountain salamanders lurking under the rocks and logs, along with Wherle's salamanders and a few other species. Thanks to a surveying error during the logging days, an area of virgin forest, with towering spruce, cherry, oak, and other trees, remains on the mountain's northeast slope.

Signs of those logging days are all over this region. During the late 1800s and early 1900s the magnificent forests cloaking West Virginia's mountains were harvested with alarming efficiency. Almost every tree was cut by about 1910, and most were transported out to the mills on temporary logging railroads. Many of these old railroad grades are today's hiking trails. Shortly after the turn of the century my grandfather was working on the logging trains outside of Richwood. Once the rich wood was gone, the Richwood mills closed down and floods ravaged the town because the barren hills couldn't hold back the water of heavy rains. When the mills closed down, the town nearly died and many left, including my grandfather, who ended up in Baltimore.

SOUTHERN MONONGAHELA

Return to the Highland Trace and continue a little farther south to enter Pocahontas County, better known than anywhere else in West Virginia as a haven for outdoor recreation. Merchants in this county have joined forces with the county's tourism board to promote the region as a destination for outdoor recreation enthusiasts. The Snowshoe Ski Resort is perhaps the county's best-known attraction, and the ski trails and snow conditions here lead many skiers to call Snowshoe the best ski resort in the mid-Atlantic region. I've never skied here, but I have visited the mountaintop on a summer evening when hermit thrush, veery, and Swainson's thrush could all be heard singing together. Cross-country skiing is also promoted at the nearby Elk River Center, which in summer becomes a center for mountain biking.

I'm not one to be lured to an area by elaborate promotions or the promise of structured outdoor sports, so I avoided Pocahontas County for many years. My objections were practical as well as philosophical, for other parts of the West Virginia Highlands are closer to my home and could be reached with a shorter drive. But I couldn't stay away forever, and my West Virginia wanderlust drew me south. When I finally explored Pocahontas County I found two particularly interesting places for nature study, the long Greenbrier River Trail and the vast region known as Cranberry Country. The Highland Trace provides access to both.

The Greenbrier River Trail follows an abandoned railroad right-of-way for 75 miles through the valley of the Greenbrier, from Cass at the north downriver to Caldwell, near White Sulphur Springs. The trail winds through a mixture of forest, old field, and farm habitat, passing through a few tiny rural communities along the way. Marlinton, the Pocahontas County seat, is right along the trail, and there's a small visitor center housed in the old train station. The trail is almost dead flat with a gravel or packed dirt surface, ideal for walking or for bicycling. Hikers and bikers see countless scenic views of the river because the trail hugs its bank for miles and crosses many old trestles that now serve as trail bridges. Many different wildflowers grow and bloom along the trail in spring and summer, with the bizarre and lovely pipevine especially abundant. Numerous springs and seeps at the trail's edge are home to many salamanders; wetlands in the river's flood-plain provide havens for many frogs. Check the rocks along the river on sunny summer days for snakes; rattlers are said to be fairly common, though I've never seen one here. I've also never found any green salamanders on the wet rocky ledges next to the trail, though I know that they must occur here. There's always something else to search for.

Although the Highland Trace leads right through the western edge of Marlinton, a better bypass route is State Route 150, known as the Highland Scenic Highway. This high-quality and scenic road is closed to commercial vehicles, is never crowded, and more than 60 percent of its 22-mile length lies above the 4,000-foot contour. There are four formal overlooks along the Highland Scenic High-way, and all are great birding spots. Listen for nesting chestnut-sided and mourning warblers at the first two overlooks. Cross the road from the third into a wet meadow and listen

Alder flycatcher on balsam fir

for alder flycatchers. Watch for the aerobatic antics of common ravens from any of the overlooks. The Highland Scenic Highway reconnects with the Highland Trace, here piggybacking with West Virginia Route 39, at the Cranberry Visitor Center, a U.S. Forest Service facility with excellent displays and helpful staff. You can pick up maps, books, and brochures here, and, at the same time, look out the windows at hummingbirds visiting the feeders. Watch the finches around the parking lot carefully; some years there's a flock of red crossbills along with the purple finches found here.

Near its midpoint Route 150 drops to its lowest spot where it crosses the Williams River. Just downstream is the Tea Creek campground, tucked comfortably into a birch and maple forest where black-throated blue warblers and veeries nest in abundance. A gravel road follows the Williams River in both directions, with many pullouts and informal camping areas along its length. The Williams is a fine fishing river, its forested banks and rippling waters providing ideal trout habitat. South from the Williams River the Highland Scenic Highway follows a high mountain that forms the eastern edge of the Cranberry River watershed. Vast unbroken forest spreads out to the west from the overlooks along this stretch of the Highland Scenic Highway.

The Cranberry River and its tributaries flow westward out of a great expanse of high country. A part of the region is preserved as the Cranberry Wilderness, another section protected as a special botanical reserve, and a big chunk is set aside as the Cranberry Backcountry. The Cranberry Wilderness, like all federally designated wilderness areas, is an area where primitive trails lead through uninhabited country, and no development or vehicles are permitted. The Cranberry Backcountry is less vigorously protected; here some of the trails are old roads where bicycles and horse-drawn carts are permitted, though both the backcountry and the wilderness are closed to logging, grazing, and mining. Both areas share the same mix of habitats ranging from high mountain ridges to rich river valleys. Along the high ridges, where red spruces and northern hardwoods dominate the forest canopy, a great assortment of northern songbirds and wildflowers can be found in late spring and summer. It looks a lot like Vermont here.

The whole Cranberry region gets a lot of precipitation, most places averaging over 60 inches (152 centimeters) annually, and the mountainslopes are covered with diverse hardwood trees and a rich understory often dominated by rhododendron. Known colloquially as *big laurel*, these rhododendrons open their big greenish-white flowers during June, a fine time to visit this area. Down in the quiet valleys of the Cranberry River and its tributaries the same type of rich forest can be observed. Scattered through the riverside forests are a few meadows kept open by floods or beaver activity. These meadows are perfect spots to visit when summer heat is at its worst; they're always cool and pleasant, and throughout the summer the meadows harbor wildflowers, butterflies, and berry bushes ripe with fruit. Watch for deer and bear visiting the meadows during the morning and evening hours when the beavers can often be seen along the creeks.

The Cranberry Botanical Area is a real gem. The area preserves four mountain bogs in a region locally known as the Cranberry Glades. A boardwalk trail leads through two of the bogs—Round Glade and Flag Glade—the other two bogs are accessible only with special permission. From a mat of cool sphagnum moss and peat grows a regionally rare plant community, highlighted by two wild orchids, grass pink and snake-mouth, that bloom in midsummer. Nesting warblers include northern waterthrush and Nashville warbler, two species at or near their southern limit of nesting range here at Cranberry. Although it is great to walk the boardwalk at any time of day, it's magic at the dawn of a summer day. I want to include a moonlight walk around the glades on my next visit to this spot.

The Highland Trace ends at Richwood, after passing close to the Falls of Hill Creek as it skirts the southern edge of the Cranberry region. Richwood's old homes still stretch up the hillsides overlooking the Cherry River. In early August the town's annual Cherry River Festival occurs; it features a parade of the Cherry River Navy, touted as the world's only Navy consisting of all admirals and no boats. I remember watching my grandfather, my dad, and my big brother marching in the parade. My grandfather grew old and weak, and we stopped going to Richwood before I grew old enough to become an admiral myself.

Along its length, the Highland Trace provides access to some of the most beautiful mountain habitat in the entire mid-Atlantic region. The combination of high altitude, cool temperatures, and abundant precipitation gives much of the West Virginia Highlands a different feel from the rest of the mid-Atlantic. These topographic and climatic factors maintain the northern hardwoods, spruce and fir groves, and beaver bogs, which are all plant communities more commonly associated with New England or Canada. Such communities contain a variety of plants and animals of the boreal zone. When summer's heat is at its worst in the urban lowlands, a trip along the Highland Trace is much like a trip to the north, with cool temperatures, northern environments, and the extra bonus of a few southern species mixed in. Yet in every season there are wonders to discover, lovely scenes to enjoy. Some day I'd like to move back to West Virginia, where some of my roots are firmly grounded. If I do, it will surely be to some place near the Highland Trace, the pathway to the richest mountain habitats in the mid-Atlantic.

7

The Potomac—
The Nation's River

Western Maryland has a strange shape. The state is just 3 miles wide near Hancock, but broadens west of there until it abruptly ends at a north-south line. As a child I learned to look at the Maryland map as if it were the outline of some strange animal, with the western portion representing the tail. The three counties of Western Maryland—Washington, Allegany, and Garrett (from east to west)—start with the letters W, A, and G; how appropriate that Maryland's tail should "wag."

History gives us a decent reason for the shape. Long before this region was explored and settled, Maryland set its boundary with Pennsylvania on a straight line that was surveyed by Mason and Dixon between 1763 and 1767.

Great blue herons along the Potomac River

For a boundary with Virginia, the Potomac River was chosen. How could a poor colonial Marylander have known that the two lines would nearly cross!

Every river has many sources, and the Potomac is fed by hundreds of tributaries in the mid-Atlantic mountains. The nominal source of the Potomac—at least its North Branch—may be found at the southwestern corner of Maryland's western tail. Surveyors marked this spot with a rock that became known as the Fairfax Stone. Today this historic site, just south of the ridgeline of Backbone Mountain where Maryland's highest elevation occurs, is protected as a park by the state of West Virginia. Surrounding the quiet, inconspicuous spring that becomes the Potomac are open meadows where, in late spring, hobomok skipper butterflies dance from yellow hawkweed flowers to orange ones. A journey down the Potomac from the Fairfax Stone, with a few side trips on important tributaries, provides an instructive transect of mid-Atlantic ecosystems. Along the way you will find many wonderful natural areas and a treasure trove of American history. Environmental history is a big part of the story too. The Potomac's water quality, once terrible because of a variety of pollutants, has been dramatically improved in the last two decades, and much of the river corridor has been preserved as a natural landscape. Let's start the journey.

Rolling mountains built of sedimentary rocks dominate the landscape around the Potomac's North Branch headwaters. Little of the land is protected here, and as a result, the region is a mixture of little farms and town, forests in various stages of growth, and large strip mines. Coal layers are often close to the surface, stacked among sandstone, shale, and limestone strata. Strip mining is the easiest way to harvest this resource, but its effect on the environment is devastating. All plants and animals at a strip mine site are destroyed and displaced, soils are severely damaged, and mine runoff is highly acidic, negatively affecting water quality in rivers and streams. Environmental legislation passed in the 1960s and the 1970s forced strip mine operators to protect surrounding waters and to restore plant communities to mine sites when work is complete. The upper Potomac's aquatic communities, once severely degraded, are slowly returning to health.

The river accumulates water from its many headwater tributaries and carves a scenic valley between the rolling green mountains of western Maryland and those of northern West Virginia. Just downriver from the town of Kitzmiller lies one of the Potomac's few reservoirs, the Jennings Randolph Lake. During the middle decades of the twentieth century the U.S. Bureau

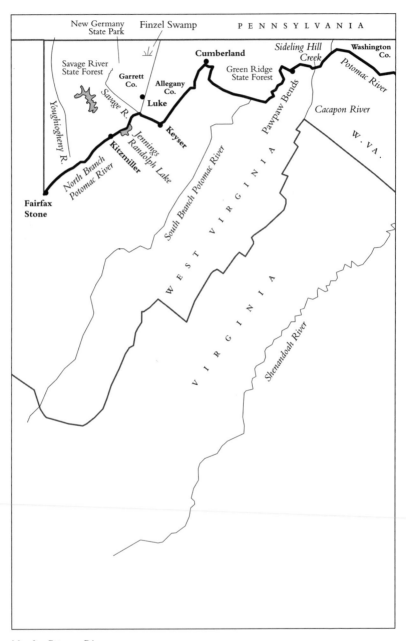

Map 3. Potomac River

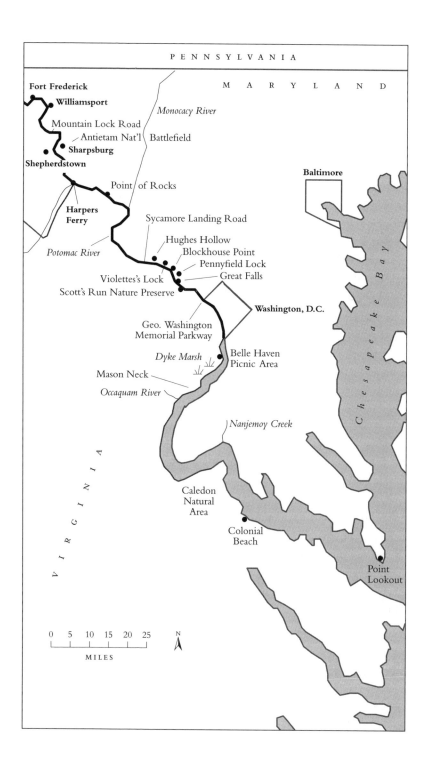

PENNSYLVANIA

MARYLAND

Fort Frederick
Williamsport

Monocacy River

Mountain Lock Road

Antietam Nat'l Battlefield

Sharpsburg

Shepherdstown

Baltimore

Point of Rocks

Harpers
Ferry

Sycamore Landing Road

Potomac River

Hughes Hollow
Blockhouse Point
Pennyfield Lock
Great Falls

Violettes's Lock

Scott's Run Nature Preserve

C h e s a p e a k e   B a y

Washington, D.C.

Geo. Washington
Memorial Parkway

Dyke Marsh

Belle Haven
Picnic Area

Mason Neck

Occaquam River

Nanjemoy Creek

V I R G I N I A

Caledon
Natural
Area

Colonial
Beach

Point
Lookout

0   5   10   15   20   25

N

MILES

of Reclamation developed a plan that would have changed the free-flowing Potomac into an engineered water system, with dozens of dams on the mainstem and its tributaries. Our region's active conservation community successfully fought against this ambitious plan during the 1950s. As a result, the river and its floodplain remain largely intact as a vibrant, dynamic landscape. The Potomac is still basically a natural river.

Natural rivers will flood from time to time. Floods can be disastrous when they cause major damage to human communities, but floods have positive effects on a river's ecosystem. Adjacent to most rivers are low-lying areas that become part of the enlarged river channel during floods. As the river waters recede from these floodplains, both organic and inorganic soil materials are left behind. Floodplains thus develop extremely fertile soils. Floodwaters sometimes carve new channels of flow, moving entire sections of a river. Abandoned channels often become ponds, marshes, or swamps, adding to the habitat diversity of the river ecosystem. Seeds and bulbs from many plants are moved downriver during all seasons, but floods deposit many of them onto the rich floodplain soils. Even terrestrial animals sometimes move downriver during a flood, floating along on vegetative rafts.

Minor floods occur about every other year along the Potomac, major floods about once a decade. I've witnessed three big floods since I began to watch the river: June 1972, November 1985, and January 1996. Roads, bridges, trails, parks, and houses were damaged by each of these floods, but the river's ecosystem benefited. I visited a favorite spot on the Potomac floodplain shortly after the 1985 floodwaters receded and was horrified to see that the flood had left a full foot of sandy soil piled on the ground. This had been a great spot for spring wildflowers, and I felt certain that this natural garden had been destroyed. I returned the next April and, to my amazement, the flowers were in full bloom, having pushed up through the new soil. I also discovered plants that I hadn't seen at this site before; the floodwaters had no doubt brought their seeds along with the new soil. One year later, after the organic flood debris had more fully composted, the spring display was better than I had ever seen before, as all the plants seemed especially robust.

The town of Luke, which lies downriver from Randolph Lake, has one of the river's larger industries, a paper plant that, when I was a kid, could easily be smelled 20 miles or more downwind. Luke also marks the confluence of the Savage River with the Potomac. The watershed of the Savage includes some of Maryland's most attractive forests, the northern hardwoods and mixed conifer communities of New Germany State Park and the Savage River State Forest. Little backroads and pleasant, seldom-walked trails lace this large forest block, a great place to visit during late spring and early summer when rhododendrons bloom and rose-breasted grosbeaks, scarlet tanagers, and more than a dozen species of northern warblers nest. Summer brings flowers to the forest edges and meadows, which are visited by butterflies found in no other part of the state, including the Harris' checkerspot and Atlantis fritillary. Fall brings lovely color to the forest, but also a small army of hunters; if you visit in autumn, first check regulations and the dates of hunting seasons. It's not a bad idea to routinely wear blaze orange on autumn outings.

Garrett County, Maryland, which includes the watersheds of the Savage and the Youghiogheny (just call it the *Yock* as the locals do), is truly the state's snow belt. Almost every winter brings a good snowpack to the region, making this a favorite area for cross-country skiing. New Germany State Park is an ideal place for beginners and experienced skiers alike, as it features groomed trails and all the amenities, but the intrepid can find dozens of other spots to enjoy the wintry scenes. I'm especially fond of clear, cold midwinter nights when the moon is full; as long as I stay away from steep hills, the moonlight reflected off the white snow provides more than enough light for skiing.

One site that deserves special attention at the head of the Savage watershed is the Finzel Swamp. Finzel is a Nature Conservancy preserve high on the western flank of Big Savage Mountain. Ruffed grouse are fairly common on the forested slopes of the mountain. Each spring, when the white clintonia and mountain laurel are in bloom, grouse perform their courtship ritual. Males move to an open spot on the forest floor or a fallen log, fluff out their neck feathers, puff out their chest, and beat furiously at the air with their wings. A low-pitched, soft drumming sound results, which apparently drives the females into a romantic frenzy.

Finzel also features a large open meadow filled with grasses, goldenrods, and other herbaceous plants. The American copper butterfly is common in summer and the rare Henslow's sparrow nests here some years. Finzel's major attraction, however, is the shrub swamp community that lines the preserve's entrance road. Big cinnamon ferns, calla lilies, and Canada burnet grow beneath tamarack trees, with a great variety of other northern plants also present. Painted trillium and Canada mayflower grow at the swamp's edge. Northern waterthrush, swamp sparrow, and alder flycatcher sing like crazy from this unusual wetland in late spring. Take a jacket to Finzel whenever you visit—this is one of our region's chilliest places.

Between the towns of Luke and Keyser the Potomac cuts a small gorge through the mountains, then beyond Keyser it turns 90 degrees to the left, winding through a broad valley between Dans Mountain on the northwest and Knobly Mountain on the southeast. Small industries, towns, and farms gradually increase downriver from Keyser to Cumberland, the first real city we reach on our Potomac voyage. Railroad yards and Interstate 68 dominate Cumberland's transportation corridor, and major thoroughfares have passed through the city for centuries. Some of the first settlers heading into the Ohio Valley traveled through Cumberland along the National Pike in the late eighteenth century. During the nineteenth century two new transportation links arrived, the Baltimore and Ohio (B & O) Railroad and the Chesapeake and Ohio (C & O) Canal. The rail line remains busy to this day, but the canal ceased operation in 1924.

Despite the lofty objective implied by the Chesapeake and Ohio Canal's name, the canal never crossed the mountains to reach the Ohio River watershed. Its total length was 184 miles (296 kilometers), running parallel to the Potomac between Washington, D.C. and Cumberland, its barges carrying coal and other products from the mountains down to tidewater. The C & O Canal and B & O Railroad were built at the same time and in direct competition with one another. Rail technology advanced rapidly, and by the late nineteenth century it was clear that the railroad was winning most of the business. Trains could move much faster than mule-drawn canal barges, and they were less labor-intensive. Most goods could be moved more quickly and at less cost by rail. The Potomac River's periodic floods damaged both the canal and the rail line, but the rail lines could be repaired more quickly and at a lower cost. As the nineteenth century closed, and a greater proportion of the forest in the Potomac's watershed was cut, flooding grew more frequent and severe. Still, the canal survived bankruptcies, floods, and declining business into the 1920s before finally going out of business.

Gradually the forgotten canal fell into disrepair, and the rich river floodplain soil along much of its length nurtured a thick, diverse young forest. By the 1950s, a wonderful forest had grown in this area, rich with wildflowers and wildlife. During this decade, a time of prosperity, growth, and development all across the nation, a new plan was hatched for the C & O Canal. A scenic highway, following the old canal right-of-way along the Potomac corridor, was proposed. Many naturalists and conservationists, having learned the value of the canal's forests and other habitats, spoke up in opposition. In 1954 Justice William O. Douglas, probably the most ardent conservationist ever to sit on the U.S. Supreme Court, invited journalists and supporters to join him on a walk of the canal's entire length. The resulting publicity and other grassroots efforts to defeat the planned highway eventually led to the creation of the C & O Canal National Historic Park in 1971. Flanking the canal's entire length from Cumberland to Washington, D.C., this park protects a series of outstanding natural areas. Cumberland is at the western end of the canal and the park; if you're ever in the mood for a delightful 184-mile hike or bike ride, find your way to Cumberland and follow the old towpath downriver. I made the entire trek on a bicycle when I was 14 years old and have been returning frequently to the canal ever since.

Cumberland lies in the heart of the mid-Atlantic's Ridge and Valley geographic province, a region characterized by long, parallel ridges that run north-northeast to south-southwest. Between the ridges are valleys, some modest and some broad, through which the Potomac's major tributaries flow. The Potomac is one of few rivers to have cut passages through the major ridges, yet it will often flow through the valleys for some distance before turning abruptly through a pass. From Cumberland east to Harpers Ferry, well over a hundred miles downriver, the Potomac alternates from a lazy, winding valley river to rocky, surging sets of rapids where the ridges are breached. During the warm weather months, this entire section of the river is popular for recreational boating and fishing. A justly popular canoeing run follows the river around the Paw Paw Bends, a series of broad, sweeping curves of the river where it flows between Town Hill, in Maryland, and Sideling Hill, in West Virginia. At times the river bangs right up against these ridges, sometimes with a gentle rocky rapid or two, but in other places it gently flows across the bottomland, with fertile farms lining its banks. The Maryland shoreline in this region is in public ownership, part of the Green Ridge State Forest.

This is dry country. The Ridge and Valley is the mid-Atlantic's driest region climatically, and the shales and sandstones underlying most of the Green Ridge forest have broken down to form quite porous soil. Oaks dominate the upland forests, while next to the rivers and streams, richer forests with a variety of trees and spring wildflowers may be found. It is on Green Ridge's harshest environments, however, where some of its most interesting communities are found, those of the shale barrens.

Shale is a sedimentary rock created by the compression of layers of mud and clay particles. Most shales are dark in color and rather easily broken. There's a lot of shale underlying the Green Ridge, and, in places the mountainslopes have become sloping piles of shale rubble. These bits of rock, rather flat and mostly between a half-inch and two inches in length, never completely stabilize on the mountainsides, but rather they slip and slide downward a little each year. Smaller bits of rock, along with organic and inorganic soil particles, tend to wash away more quickly, leaving behind just the rockpiles. Thus the shale slopes are mobile rockpiles, not exactly the ideal situation for most plants. Not too many species are adapted to grow on the shale slopes, leaving a rather open environment, known as a shale barren.

Botanists give a more precise definition to the term shale barren, referring specifically to certain outcrops of the mobile shale hillsides found on south-facing slopes. These slopes directly catch the intense sunlight of midday, and the dark-colored shale stones absorb a lot of heat. As a result, in summer the south-facing mobile shale hillsides act like solar ovens, heating up to quite high temperatures. As noted previously, there is little soil on these slopes, and thus little moisture retention; shale barrens aren't just hot, without soil, and mobile, they are also extremely dry. Few plants have adapted to grow on these sites, but among those which have are a group of rare plants called shale barren endemics, which exist only in this habitat and nowhere else in the world. Explore the backroads of the Green Ridge State Forest and you'll find plenty of open shale slopes, some harboring shale barren endemics such as Kates Mountain clover, shale barren evening-primrose, and mountain pimpernel.

I love to poke along the backroads of Green Ridge. Dirt roads wind through the hills and valleys, and many junctions are poorly marked (if they are marked at all), so it's easy to get a little lost. Watch the landscape, though, remembering the pattern of the Ridge and Valley's parallel mountains, and it's impossible to become hopelessly disoriented. Okay . . . if you're less adventuresome you can always take a map. Sometimes I've gone by car, stopping here and there to explore, but I've also enjoyed bicycling through Green Ridge (though some of the uphill sections were pretty challenging!). In early summer chestnut-sided warblers and prairie warblers sing along the ridgetops, with Kentucky warblers and black-throated blues on the wooded slopes. Patches of milkweed, including the orange-flowering one called butterfly weed, are found along the roadside in many places, and when they bloom in midsummer they are great places to find butterflies. Aphrodite fritillary, northern metalmark, and American copper are sometimes quite common at Green Ridge.

Half a dozen miles west of Paw Paw, the South Branch of the Potomac flows north from West Virginia into the North Branch. The South Branch's headwaters include some of the mid-Atlantic's highest mountains (see West Virginia chapter for details about this region). Downstream from Green Ridge are three more tributaries well worth exploring—Sideling Hill Creek in Maryland and the Cacapon River and Sleepy Creek in West Virginia. Each has extensive floodplain forest along its banks and a vibrant aquatic ecosystem. Come late summer, the extremely rare plant harperella may be found with its inconspicuous flowers in bloom along the shallows or wet banks of both Sideling Hill Creek and Sleepy Creek.

The Potomac River continues to meander through valleys and past ridges, swinging northeast to Hancock where it almost intersects the Mason-Dixon line. From here the river's course switches back to the southeast, passing close to Fort Frederick. This Maryland State Park has a historical theme, depicting the eras of the French and Indian War, the American Revolution, and the Civil War, yet its natural areas are also a worthy attraction. Woods and fields in the park and the adjacent long pondlike section of the C & O Canal called Big Pool are interesting to visit in spring and summer when a great variety of wildflowers can be found along Fort Frederick's wooded trails. Bluebirds are abundant year-round. As at many parks, large sections of lawn here

Basking painted turtles

have been allowed to revert to less intensively managed meadows, which are mowed just once or twice a year. These meadows fill with summer wild-flowers and their myriad insect pollinators, including many showy butterflies. Look closely at goldenrod flowers during late summer and you will find vivid yellow- and black-striped long-horned beetles. The still waters of Big Pool are an ideal place to search for turtles, frogs, and water snakes. Beavers are also active at Big Pool and other nearby ponds. On sunny summer days it's nearly impossible to count the numbers of painted turtles basking around Big Pool. But bring your insect repellent in summer because mosquito larvae also thrive in the quiet pond waters. Access to Fort Frederick, Big Pool, and this section of the C & O Canal is from Maryland Route 56, just south of Interstate 70.

Downstream from Fort Frederick the Potomac opens into the Great Val-ley, known in Virginia as the Shenandoah Valley, in Maryland as the Hager-stown Valley. The pleasant towns of Williamsport and Sharpsburg in Mary-land, and Shepherdstown in West Virginia, lie near the river in the Great Valley, as it winds back and forth past fertile farm fields and pastures on a generally southward course. Just north of Sharpsburg is the Antietam Na-tional Battlefield, site of the bloodiest single day's battle of the Civil War. To-day this site is peaceful and serene, perhaps the most fitting memorial to those who lost their lives in this 1862 battle. It's not a bad place for nature study, either, especially along Antietam Creek, where spring wildflower dis-plays compete with migrant warblers for a naturalist's attention during late April and May.

South of Sharpsburg the Potomac reaches Shepherdstown, where it turns to the east and bumps into the Blue Ridge, then winds back to the south to join up with the Shenandoah River before cutting across that formidable ridge. Much of this section is underlain by layers of limestone rocks, which were formed by the compression of shells that accumulated under shallow seas such as those of today's Gulf of Mexico. Such a sea covered the mid-Atlantic region for extended periods during the Paleozoic, long before the Appalachian Mountains were formed. Limestone dissolves readily in the presence of mild acids and breaks down into highly basic soils. Consequently, the soils of limestone regions support a variety of otherwise infrequently occurring plants, including walking fern, purple cliffbrake, rock sandwort, and shooting-star. You can visit excellent examples of this habitat along the C & O Canal, near its access point at Mountain Lock Road.

Extensive cavern systems often develop in limestone regions, and this is certainly true in the Ridge and Valley region of Maryland, Virginia, and West Virginia. A surprisingly large assortment of animals exist only in these caves, some known from just a single location—there are species of fish, salamanders, and various invertebrates that can only exist in caves. Many bats, including the endangered Indiana bat, roost in caves during the day, and some hibernate underground through the winter. Other animals sometimes find shelter near a cave entrance. We know that this has been true for a long time because many skeletal remains of saber toothed cats and other prehistoric animals have been found by paleontologists studying the subterranean mid-Atlantic limestone belt.

Purple cliffbrake

Caves are especially numerous in the limestone regions throughout the Great Valley. Substantial knowledge and proper equipment are needed to explore wild caves, however, and most caves that harbor rare species are closed to public entry. Commercial tours make it easy to visit some caves, but most tours rarely focus on the underground living community. I leave the study of this ecosystem to the experts and to the members of the avid caving community.

The Shenandoah and Potomac meet in a dramatic gorge cut through the Blue Ridge by the dual force of these two important rivers. The town of Harpers Ferry sits right on the bank of both rivers at their meeting point. I think it's one of the region's most beautiful places. The town has a storied history, its best-known event being John Brown's raid on the Harpers Ferry armory in an attempt to arm freed slaves just before the beginning of the Civil War. The old downtown, that region closest to the rivers, is preserved by the National Park Service in nineteenth-century style. Historical demonstrations are held here throughout the year. I especially recommend the ghost tour, which is held after dark and is very scary indeed. Trails, including the venerable Appalachian Trail (see Appalachian Trail chapter for more details), lead through oak-dominated forests up to extraordinary vista points on the surrounding mountains. Views are nice on the riverside trails too, where the diversity of wildflowers, trees, and songbirds is much greater than that of the ridges. The rapids of both rivers make a fun whitewater trip for experienced canoeists and kayakers, and bass fishing is often good near the confluence.

Just east of Harpers Ferry, the Potomac bumps through one more mountain, the Catoctin, at a gap known as Point of Rocks. The core of the Catoctin ridge is primarily greenstone, a sturdy metamorphosed basalt. The Potomac has managed to cut little of the mountain away outside of the river's channel, leaving almost no floodplain and barely enough room for both the railroad and the canal.

The Potomac next settles into a lazy stretch through the western Piedmont. One favorite spot of mine along this section is Hughes Hollow, an informal place name referring to the C & O Canal access at Sycamore Landing and the adjacent McKee-Beshers Wildlife Management Area, which is owned and managed by the state of Maryland. Gated dirt roads and semiformal trails crisscross the area, providing access to an excellent variety of fertile habitats. This is a great place to explore in all seasons, though I'm cautious about autumn visits when the McKee-Beshers section is open to hunting. The most dangerous time to visit is during the deer season, which usually runs for about two weeks near Thanksgiving. In the spring there is also a short hunting season for wild turkey.

Downriver from Sycamore Landing the C & O Canal towpath traverses a broad section of floodplain where soils are exceptionally fertile. A mature forest with many big trees has developed along this stretch. Walnuts, sycamores, ashes, elms, maples, and other species that can tolerate periodic flooding dominate this forest; most oaks and hickories, trees that are common in many mid-Atlantic forests, can't tolerate saturated soils and are scarce here. Migrant thrushes abound during May and September along the old canal bed, which is often dotted with frog-filled puddles. Away from the river, Sycamore Landing Road passes through forests that are variably swampy during wet years and big open meadows that become overgrown tangles by midsummer. Watch for eastern cottontail rabbits and red foxes here; both seem to thrive at Hughes Hollow. Observe the nesting boxes in the meadows for American kestrel (in the big boxes), tree swallow (small boxes), and eastern bluebird (medium boxes). Watch and listen for common snipe and American woodcock in these fields during late winter and early spring. Check the meadow flowers throughout the warmer months for butterflies; great spangled fritillaries swarm around milkweed blossoms in midsummer, and, if you're watchful and lucky, you might spot a silvery checkerspot or a harvester here.

The primary access points to the McKee-Beshers section of Hughes Hollow lie along Hunting Quarter Road, a lazy dirt track that parallels River Road and the Potomac for a couple of miles. A trail near the western end of Hunting Quarter Road, near its junction with Hughes Road, leads past several shallow freshwater impoundments. This is an enchanting place; I've probably been here more than a hundred times, yet I always find some surprise. The first pools are filled with spatterdock, cattails, pickerelweed, arrow arum, and other emergent plants, including the fragrant and relatively uncommon calamus, which is also called sweet flag. Wood ducks, blue-winged teal, ring-necked ducks, Canada geese, and other waterfowl are often found here fall through spring, except in cold winters when the water freezes. In summer great blue herons fish from open spots while green herons and least bitterns lurk behind the vegetation. During summer's warmest days, dragonflies dart over the water in wild feeding and mating frenzies, while northern water snakes stretch out placidly on sunny banks and low shrubs. Visit these ponds in late March or April on a cool, moonlit evening and get ready for noise; frogs thrive at Hughes Hollow, and the din of a thousand spring peepers, accented by green frogs, wood frogs, bullfrogs, pickerel frogs, and American toads, can just about drive you mad. Barred owls will probably be calling in the distance, but on the froggiest nights they can barely be heard over the anuran roar.

The second pool on the right is partially a wooded swamp, and here you may find beavers, black willows, swamp sparrows, yellow warblers, and willow flycatchers. Beyond the ponds lies an intriguing habitat mix, including old fields, actively farmed fields, upland deciduous woods, swamp forest, a spruce grove, a pine grove, and broad, bird-filled hedgerows. I've found several fox dens back here and have seen more kinds of birds, butterflies, and wildflowers than I can remember. Eventually this trail winds to the C & O Canal at a point about a mile east of Sycamore Landing. Except in wet seasons (when the canal can be wet, muddy, and uncomfortable to cross), a great circuit walk leads down this trail to the canal. You may then walk west to

Sycamore Landing, back Sycamore Landing Road to River Road, and east along River Road a short stretch to Hughes Road, which quickly turns into Hunting Quarter Road just a hundred yards from the main parking area. It's just a 3- or 4-mile walk, but it usually takes me 5 hours or more, there is so much to stop and see. Farther east along Hunting Quarter Road are several other trails that always beckon and never disappoint. The road itself has little traffic and makes an enjoyable, if sometimes muddy, walk. Watch for butterflies such as the hackberry, tawny emperor, and American snout near the hackberry trees that line the road.

Follow the Potomac's flow a few more miles to reach another transition at Blockhouse Point. Here the soft rocks and sediments of the western Piedmont are replaced by the hard, metamorphic rocks of the Fall Line. The river tumbles across Seneca Rapids at Blockhouse Point, a beautiful scene when viewed from the C & O Canal between Violette's Lock and Pennyfield Lock. Watch for common mergansers bobbing through the rapids in early spring. Between Seneca Rapids and Great Falls are many islands that can be reached by canoe. These islands have the rich soil, big trees, and profuse spring wildflower displays of the river's floodplain. Elm Island, at the base of Seneca Rapids, offers an especially nice floral display.

Another viewpoint of the Seneca Rapids can be reached along trails that lead to the high bluffs of Blockhouse Point Park, accessed from Maryland Route 190, River Road. This county park features trails that lead through extensive, dry, oak-dominated woodlands, past a few little creeks, to a high bluff overlooking the river and the canal. The park's trails are never crowded, and the diversity of its flora and fauna is quite high. Breeding birds that require large areas of unbroken forest, such as worm-eating warbler and hooded warbler, nest here in abundance. Follow the main trail along a pipeline clearing past more than twenty kinds of trees between the parking area and the canal. Look carefully at the ground next to the bluffs above the river and you may find charred remains of the old Civil War-era ammunitions caches, called blockhouses, which were built here as part of the Union's defense of Washington, D.C..

For the next 20 miles, from Blockhouse Point to the tidal basin in Washington, D.C., the Potomac alternates between quiet channels dotted with islands and churning sections of rapids and waterfalls. Lining the river are exceptionally rich floodplain forests, where the trees grow to impressive size and spring wildflowers carpet the forest floor. Spring beauty, Virginia bluebells, toadshade trillium, and many other species make the Piedmont Potomac floodplain a spectacular natural flower garden from late March through early May. Although many of these flowers are widespread, most are more vigorous and abundant here than in other areas. Uncommon plants may also be found in the floodplain, including species whose ranges are primarily to our west. Harbinger-of-spring, wild leek (known as *ramps* in Appalachian vernacular), and Coville's phacelia are among the wildflowers with disjunct populations along the Potomac.

Millions of people live and work within a few miles of this stretch of the river, yet the area remains surprisingly wild and beautiful. This is one reason I believe that nature lovers who live near Washington, D.C. are more fortunate than those in other eastern cities: Our major river, the Potomac, and its surrounding woodlands are protected as a natural corridor in both the city and the suburbs. Rivers adjacent to other large cities have become sites for major industrial development. Also, as mentioned previously, because there isn't a major dam system to control the Potomac's flow, the river periodically floods and organic nutrients are regularly added to the floodplain soils. The floods have also discouraged human settlement and development in the areas adjacent to the river, making their acquisition as parks possible.

Wildlife is abundant along the Potomac floodplain partly because of basic ecological energetics. A rich ecosystem occurs in an area where a large amount of the sun's energy is used by plants for growth, and where plants grow profusely there is a plentiful food supply for the entire ecosystem. Raccoons, foxes, skunks, red-backed salamanders, painted turtles, and other animals can only be abundant in areas where they have lots of food. The floodplain forest also supports rich wildlife communities because protection provides a long, continuous corridor of forest. Such unbroken corridors are especially important for mobile animals such as migratory birds. Birders flock to parks along the river every spring to search for migrant songbirds. Because the area's protected habitats also support a diversity of nesting species, the lucky birder who visits the river on the day of a migration wave may be able to find more than twenty species of warblers in a morning, along with several different thrushes, vireos, flycatchers, orioles, and more. Because there are many old trees and rotting snags, woodpeckers are abundant in the riverside forests. Organized bird counts are held all across the country in late December, and counts along the Piedmont Potomac often boast the nation's highest totals of pileated, red-bellied, and downy woodpeckers.

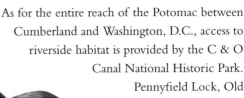

As for the entire reach of the Potomac between Cumberland and Washington, D.C., access to riverside habitat is provided by the C & O Canal National Historic Park. Pennyfield Lock, Old Angler's Inn, Carderock, Georgetown, and especially Great Falls are busy places for outdoor recreation in addition to nature study. On warm weekends the trails fill with fitness walkers, joggers, and bicyclists, the latter's numbers increasing rapidly in recent years with the development of rugged

Pileated woodpecker

mountain bikes. These are days when the canal parking lots completely fill and a nature walk along the towpath can be hazardous. If you stop to look at birds or turtles be sure to step to the edge of the trail, lest you get run over. There's now a 15 miles per hour speed limit for bicycles on the towpath.

Great Falls, the most impressive waterfall on the Potomac, is about a dozen miles upriver from Washington, D.C.. National park land protects both shores of the river around the falls. Great Falls Park lies in Virginia, and the Maryland bank is part of the C & O Canal Park. On either shore there are excellent trail systems that lead through habitats ranging from floodplain forests to rocky riverside bluffs and to scenic views of the falls.

Those views are easier to reach in the Virginia park, where short trails lead from the picnic area to the overlooks. Other trails follow the river below the falls along Mather Gorge, where vertical rock cliffs drop dramatically almost 100 feet from trail's edge to the river. These cliffs are popular rock-climbing spots, and the rapidly flowing river in the gorge is a kayaking playground during the warmer months. Peregrine falcons once nested on these cliffs, and migrant raptors are commonly seen over the gorge in autumn. Swifts and swallows dart over the river from spring through fall.

Most of the other trails at Great Falls, Virginia, wind through dry upland forests. The Swamp Trail is an exception, following an abandoned channel of the Potomac through what is now a low area where swamp forest grows. Red maple, willow oak, black gum, and spicebush are among the most common woody plants of this swamp. Songbird watching is great in both the swamp and upland forest, with hooded, Kentucky, worm-eating, and cerulean warblers among the nesting species. Botanists find many interesting plants here, some highlights being fringe-trees that grow on the bluffs, floodplain gardens with masses of blue phlox and golden alexanders, and a few large patches of horsetails near the river's edge. Along Difficult Run, at the park's southern boundary, a trail leads along dry hillsides where flowering plants include trailing arbutus, yellow stargrass, and rattlesnake weed. Butterfly diversity is high throughout the park, with Juvenal's duskywing incredibly common during April and May and the regionally uncommon northern pearly eye fairly easy to find in the moist bottomland forests from June through August.

A narrow channel of the Potomac separates the Maryland shore of Great Falls from a group of sizeable islands. Historical flows of the river and recent floods have periodically swept over the islands, preventing much soil from developing on top of the rock. The result is an environment known as bedrock terrace, which has low soil fertility and is extremely hot and dry during summer. The forest cover is open and thin, with drought-tolerant species such as scrub pine and post oak struggling to reach modest height. Many open meadows intertwine with these wooded patches, and here grow plants more common in the prairies of the Midwest, such as blue false indigo. The trail that leads to Maryland's best view of the falls crosses Olmstead and Falls Islands and leads through fine examples of bedrock terrace habitat. Access to this trail was abruptly cut off in June 1972 when floodwaters tore away the foot bridges. Twenty years later the bridges were rebuilt, along with a boardwalk trail designed to keep visitors from inadvertently trampling across rare plant communities. In January 1996 floods damaged the bridges and boardwalk, once again closing access to the islands.

The Billy Goat Trail, just downriver from Great Falls, follows the Maryland shore of Mather Gorge for about 2 miles. This rocky trail begins and ends at the C & O Canal towpath, leading through floodplain and swamp forests to the high bluffs overlooking the river. The gorge was cut by the river's falls, which for centuries have been migrating upstream as they continuously erode through the bedrock. This is an area where the rocks you see today were twisted, fractured, heated, and compressed during the creation of the Appalachian Mountains. Dominant rocks are schist, which is a metamorphosed shale, and metagraywacke, a metamorphosed sandstone. Both are very hard and resistant to erosion. High waterfalls are almost always formed in hard rocks; soft ones would erode quickly from dramatic falls to more gradual cascades.

The Billy Goat Trail travels over extensive rock outcrops where these two metamorphic rocks can be closely viewed and where broad bands of obviously different ones can also be observed. Molten rock squeezed between fragmented blocks of the schist and metagraywacke during the calamitous time of the Appalachian uplift, cooling to form dikes and irregular blocks of igneous granite, amphibolite, and lamprophyre that intermix with the metamorphics. One major fault that later cracked this matrix became the soft area eroded by the river through Mather Gorge. There's clear evidence on the top of the bluff that before the gorge was cut the river flowed across this broad, rocky terrace: Rounded potholes, some with river-rounded pebbles in the bottom, are quite common along certain sections of the Billy Goat Trail. Potholes are formed at the base of river cascades by swirling waters and the rocks and pebbles moved along with the water.

Several other parks line the Virginia shore of the Potomac River west of Washington, D.C. and provide access to the fertile habitats of our lower Piedmont similar to those of the C & O Canal and of Great Falls Park. Riverbend Park is just upriver from Great Falls, and although the area doesn't have falls or a rocky gorge, it still supports a mix of habitats. It's a good spot to look for birds, wildflowers, animal tracks, or to simply walk—foot trails provide access to beautiful forests and the river's edge. In summer be sure to check the big meadow near the Nature Center for butterflies.

Scott's Run Nature Preserve is downriver from Great Falls, just outside the Beltway. Here, too, fine trails lead through rich forests. The little gorge cut by Scott's Run as it reaches the Potomac is a scenic gem. Some of our region's most mature upland woods grow on the hills east of the creek. Wildflowers that depend on undisturbed, mature forests have disappeared from most areas near Washington, D.C., but many still grow at Scott's Run. Among these sensitive species are wild orchids such as yellow ladyslipper, puttyroot, and large whorled pogonia.

The idea of the George Washington Memorial Parkway seems incongruous at first—a national park that is a commuter highway. The land of the parkway is a corridor that includes the road, but this park also protects an excellent forest along the Potomac shoreline between the Beltway and Rosslyn. The easiest access is from the Turkey Run picnic area, where trails lead steeply down the river bluff to a floodplain that supports an extraordinary plant community. By mid-April the show of Virginia bluebells here is stunning, with thousands of these showy plants and their sky-blue flowers lining the path. Look carefully here and you may find white trout-lilies, false mermaids, ramps, twinleaf, and many other regionally uncommon flowers. Watch for wood ducks in the quiet river channels; most likely you won't notice them until they burst into flight, giving their squeaky, high-pitched calls. Pileated woodpeckers are common here, and each little creek in this area hosts a few nesting pairs of Louisiana waterthrushes. Palm warblers pass through here during migration, their numbers usually peaking around the best wildflower time in mid-April, with other migrant songbirds most abundant in early May.

Downriver from Washington, D.C. the Potomac River takes on a very different character. Like many cities along the Eastern Seaboard, Washington, D.C. was developed along the Fall Line, where rivers reach sea level (or *tidewater* as you'll often hear it called) and become navigable. From Washington, D.C. to the Potomac's mouth at the Chesapeake Bay, the river is broad, gently flowing, and often lined with marshes. Just south of Alexandria, Virginia, which lies directly across the river from Washington, D.C., the Beltway's Woodrow Wilson Bridge crosses the river. South of the bridge, along the Mount Vernon Parkway, lies the Belle Haven picnic area. Submerged aquatic plants flourish in the broad, quiet stretch of the river adjacent to the picnic area. By late summer the floating vegetative mat looks like an island. Bass flourish in this section of the river, which also attracts a great variety of birds. Watch for shorebirds, herons, and egrets on top of the mat from midsummer through autumn. As the plants die back after fall's first freezing nights, large numbers of ducks, Canada geese, double-crested cormorants, and tundra swans arrive from nesting areas to the north. Dozens of bald eagles patrol the entire Potomac from Washington, D.C. to the Chesapeake during the colder months. Large flocks of gulls also gather along the river at the same time. Most are ring-billed and herring gulls, our most common species, but sometimes regional rarities such as Iceland gull or lesser black-backed gull may be found in the flocks. Visit Belle Haven on any weekend morning and you'll find many birders gazing intently over the river.

Dyke Marsh is an extensive freshwater tidal wetland immediately south of Belle Haven. Sparrows and other songbirds lurk in the brushy woods that line the marsh trail, which offers great views of this extensive wetland and of several broad stretches of the Potomac. Many of the wetland plants bloom during summer, with the showy mallows among my favorites. Birding is excellent in all seasons at the many habitats of Dyke Marsh.

Mason Neck is a peninsula jutting into the river along the Virginia shore near the mouth of the Occoquan River, about 15 miles south of Washington, D.C.; most of the land here is protected within Mason Neck State Park and the Mason Neck National Wildlife Refuge. Here you can find several scenic trails that lead through extensive forests or to wetlands. The young forest here has a Coastal Plain character, with many beech and sweetgum trees in the canopy and American holly as a common understory component. These woods, swampy in places, are home to white-tailed deer, barred owls, many frogs, salamanders, and a good variety of songbirds.

The wetlands at Mason Neck are superb wildlife habitats, particularly the Great Marsh, a very large freshwater tidal marsh within the national wildlife refuge. The Wood Marsh and Great Marsh Trails lead to the edge of the marsh, which is extensively used by waterfowl in winter and during migration. Northern harriers, swamp sparrows, common snipe, and many other birds use the marsh. Blackbirds are abundant here; a few years ago a red-winged blackbird with an aberrantly colored pure white tail wintered in the marsh. A few bald eagles nest at Mason Neck and many more overwinter here, with the trees at the edge of the Great Marsh a favorite roosting spot. All the peninsula's habitats can provide excellent birding during migration.

From Mason Neck to the Potomac's mouth, where it empties into the Chesapeake Bay, the river is brackish and, in truth, could be considered an arm of the Chesapeake. Boats ply its waters, everything from canoes and small sailboats to big commercial barges. Extensive forests and wetlands line the banks of the Potomac and its many lazy Coastal Plain tributaries, along with scattered crop fields and pastures.

A favorite Potomac tributary of mine along the river's tidal stretch is Nanjemoy Creek, which lies southwest of La Plata in Charles County, Maryland. Like other streams of the Coastal Plain, Nanjemoy has two distinct sections. At its head, the creek gently flows through rich woods and quiet farm country. Downstream, where the creek reaches sea level, it broadens to wide, bay-like dimensions, lined by extensive marshes. Tides push the water back and forth through this section, bringing just a bit of salt water into the wetlands, so the habitat is that of brackish marsh. It's a wonderful waterway to explore with a canoe or other small boat. A public boat launch is accessed from Route 425 at Friendship Landing. Poke around the wetlands north of the landing and you'll find areas frequented by beavers, red-headed woodpeckers, and bald eagles. Great blue herons are always numerous here; the region's largest rookery of this colonial-nesting species is upstream from the landing along Nanjemoy Creek, its mature forest protected as a Nature Conservancy Preserve.

Caledon Natural Area, a Virginia state park, lies across the Potomac from the mouth of Nanjemoy Creek. Access to the river is limited here to minimize negative effects on bald eagles, which use the park's big riverside trees as roosting sites, especially in summer. Park rangers offer guided visits to the roost sites. Caledon also offers lovely, open-access trails through its pleasant mix of forests and fields; just watch out for chiggers in the grassy areas near the visitor center.

Follow the Virginia shore downriver from Caledon to reach the small town of Colonial Beach. Earlier this century the town thrived as a small resort, in part for a quirky reason. Charles County, Maryland, permitted limited legal gambling for many years, right up to the 1960s. Virginia, though, had strict laws against gambling. Surprisingly, the boundary between the two states is drawn along the Virginia shoreline rather than through the middle of the river. Clever entrepreneurs in Colonial Beach realized that a building placed on pilings out in the river was technically in Maryland, and gaming was thus permitted. Times have changed and Colonial Beach is now a quiet place, a great spot for scanning the broad waters of the Potomac in late autumn, when loons, grebes, and diving ducks can be plentiful.

There's one more excellent natural history destination to include in this discussion of the Potomac, a spot that every fan of the river should visit: Point Lookout. Located on the Maryland side of the Potomac's mouth, Point Lookout State Park is a mix of forest, field, wetland, and vast expanses of water. All this water moderates the climate of Point Lookout, keeping it warmer than nearby areas in winter and cooler in summer. The combination of the point's warmer winter weather and its location, 85 miles south of Baltimore, gives the area a distinctly southern flavor. Some plants and animals that are common in the southeastern United States are near the northern edge of their range here. Loblolly pine forest covers much of the park, with brown-headed nuthatches and pine warblers nesting in these groves. Coastal Plain hardwoods, such as white oak and sweetgum, dominate other groves. Head out toward the point, however, and the trees gradually give way to open fields and small shrubby patches.

The outermost point is fenced and closed to the public. The edge of land near the point alternates between brackish marshes and sandy beaches. The west side lines the Potomac, 8 miles (13 kilometers) wide here at its mouth. To the east lies the Chesapeake Bay, which is over 15 miles (24 kilometers) wide here. These broad open waters provide ideal habitat during the colder months for common loon, horned grebe, and many different diving ducks, such as oldsquaw, bufflehead, and common goldeneye. During fall migration, the point acts as a funnel for southbound songbirds, raptors, and monarch butterflies, all of which would rather fly over land than over water on their long journeys. The mix of migration's vitality with the presence of southern plants and animals makes Point Lookout a rewarding place to visit in autumn. Butterfly enthusiasts visiting Point Lookout in September can often find clouded skipper, ocola skipper, long-tailed skipper, white-M hairstreak, cloudless sulphur, and other species more typical of areas farther south.

The Potomac River, called *the nation's river* in some history books, has escaped many of the plagues afflicting eastern rivers. No major dam system tames its flow. Few major industries occur on its tributaries or main branch. Many protected natural areas lie along its length. For much of my life, the Potomac has been one of my most important teachers. Sometime I'd love to take a year to spend along the river, checking out little crannies and corners I've never explored in the river, its surrounding forest, field, and wetland habitats.

8

Ridge and Valley Region— Appalachian Trail Country

Long before I ever watched birds, keyed out wildflowers, or chased butter-flies, my interest in nature was nurtured by a love of hiking. My first hikes, as an early teen, kept me close to home, as I'd visit places like the C & O Canal, Sugarloaf Mountain, and Seneca Creek State Park. As I gained a lit-tle stamina and know-how I set my sights higher. I wanted to walk along what I considered to be the ultimate hiker's path: the Appalachian Trail. I didn't know much about the trail, but its name was always spoken with reverence. I knew it led to some of the most spectacular mountains in the Appalachians, and I knew that it was a very long trail.

Before I turned sixteen I managed to set foot on this storied path, with initial weekend forays to the Michaux State Forest in southern Pennsylvania,

Hawks migrating over ridge

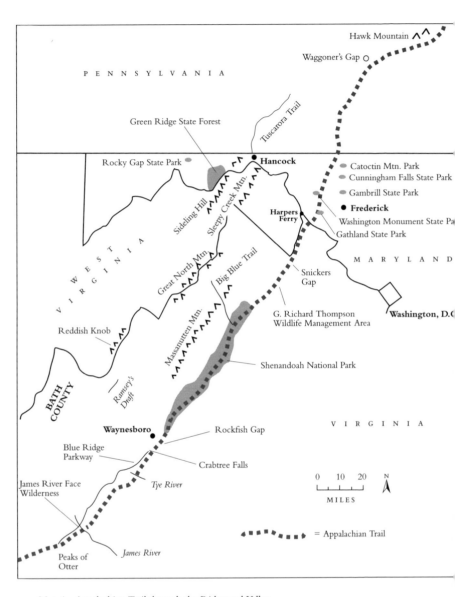

Map 4   Appalachian Trail through the Ridge and Valley

to Annapolis Rocks and Black Rock in Maryland, and to Shenandoah National Park in Virginia. As my strength and experience grew, so did my ambition. I managed several hikes of four, five, and even seven days before I finished high school. I read everything I could find about the trail, joining the Appalachian Trail Conference and attending its annual conference at age seventeen. By college I was regularly hiking 20 miles a day, even meeting the bizarre challenge of the University of Maryland's Terrapin Trail Club by hiking the state's entire stretch of the Appalachian Trail, about 39 miles (63 kilometers), in a single day. I dreamt of the day I could carve out four or five months from my life and hike the entire 2,100-mile (3,400-meter) length of this legendary trail.

I still haven't done it. I've never managed to find the time but, in truth, the Appalachian Trail and the other hiking destinations of those earlier years changed my plans. You see, it was while I wound through some of the most beautiful mountains of the East that I started exploring natural history. First I wanted to put names to the wildflowers I was seeing. Then it was the trees, the ferns, and the shrubs. Before long I was watching the birds and all the other critters I could find, from salamanders to butterflies. I was changing from a hiker who liked nature to a naturalist who liked to hike. Slowly the miles I'd cover in a day decreased, while my satisfaction with each mile traveled increased. The Appalachian Trail gave me an avocation, which turned into a career, which matured to become the passion of my life, the study of this wonderful planet. Am I grateful to the Appalachian Trail!

The Appalachian Trail follows from ridge to peak to mountain stream all the way from Springer Mountain, Georgia, to Mt. Katahdin, Maine. In the mid-Atlantic, the Appalachians divide neatly into two regions: the Allegheny Plateau at the west and the Ridge and Valley to the east. The Appalachian Trail keeps to the east, sticking mostly to the Blue Ridge, our very easternmost major mountain chain, and always staying well within the Ridge and Valley region. An overview of this venerable footpath, along with associated side trails, takes us to some of the best natural areas in the Ridge and Valley. If you're really ambitious, you can follow along on foot! Don't worry if you're not a hiker, though, as most areas can be reached by car and by short strolls. Let's head from south to north.

The Appalachian Trail stretches for hundreds of miles across the high forested ridges and open mountaintop balds of the south, passing through Georgia's green and rolling mountains, Carolina's Nanatahalas and Smokies, and the hypnotic long ridges of southern Virginia. We'll pick up the trail south of the towns of Lexington and Buena (pronounced *Bew-na*) Vista, about a hundred miles west of Richmond. Here Virginia's mightiest river, the James, has cut a fantastic gorge through the Blue Ridge. A small but biologically significant federal wilderness area, called the James River Face, preserves the forests and cliffs of this dramatic gorge. You can start a walk along the Appalachian Trail down at the river, but be prepared for a long uphill walk because the surrounding mountains are 2,000 to 3,000 feet higher.

Highlands near the Appalachian Trail in this region are easily accessed, however, from the Blue Ridge Parkway. This scenic road covers almost 500 miles through North Carolina and Virginia. The road and a surrounding corridor of protected land are managed by the National Park Service, protecting a great assortment of mountain environments between Great Smoky Mountains National Park, the road's southern terminus, and Shenandoah National Park at the north. The Blue Ridge Parkway never leads into the James River Face Wilderness (by definition wilderness areas must be roadless), but it does lead through similar habitats. One can spend many spring, summer, or autumn days happily poking along the Blue Ridge Parkway, enjoying short sorties through quiet forests and peaceful meadows. Winter drives along the parkway can be beautiful when snowstorms haven't closed the road. All along the parkway are mountainside vistas framing mosaics of dense green forest and pale green pastures, with little farms and cottages scattered throughout. I like to listen to Beethoven's Pastoral Symphony when traveling through this stretch, though local radio stations are more likely to play the songs of Lester Flatt, Doc Watson, Bill Monroe, and other bluegrass musicians.

The Blue Ridge Parkway and the Appalachian Trail follow the same mountains through much of southern Virginia. One regional highlight that the parkway accesses but the trail does not is the Peaks of Otter, which lies 20 miles south of the James River Face. The high, blocky mountains here are home to stunning wildflower displays in spring and early summer, and an endemic amphibian, the Peaks of Otter salamander. A number of salamanders have limited ranges within the southern Appalachian Mountains. Species of the northern conifer forest habitat spread across this region during the last Ice Age, when the boreal forest was widespread. As the climate warmed, this habitat became restricted to the highest peaks, and many of the salamanders that depended on this type of forest were isolated and thus evolved into distinct species. Check under rocks and moist logs high on the Peaks of Otter and perhaps you'll find its namesake salamander.

The northernmost stretch of the Blue Ridge Parkway and its paralleling section of Appalachian Trail run through the Pedlar district of the George

Washington National Forest. Crabtree Falls, one of the highest waterfalls in Virginia, tumbles down from the slopes of Pinnacle Ridge just a few miles away from either thoroughfare. This dramatic cascade is especially impressive at times of high water, which generally come after warm rains melt mountain snow in spring and after heavy summer thundershowers. My favorite section of the Appalachian Trail in the Pedlar is the crossing of the Tye River, whose gorge lies more than 3,000 feet (900 meters) below its flanking peaks, Priest Mountain to the south and Three Ridges to the north.

West of the Blue Ridge in central Virginia, a broad section of the Ridge and Valley region is within the boundaries of the George Washington National Forest. I enjoy exploring the areas west of Staunton and Harrisonburg, where many hiking trails and backroads lead through rich forests and up sturdy ridges, the highest more than 4,000 feet (1,200 meters) in elevation. You can drive to the summit of Reddish Knob; views from this lofty peak stretch all the way across the Ridge and Valley, from Spruce Knob on the Allegheny Plateau to Shenandoah Park and the Blue Ridge. Forest Road 85 winds near the summit of Shenandoah Mountain north and south of Reddish Knob, leading through scraggly oak forests where ruffed grouse and wild turkey are common. A short trail to Bother Knob begins near the northern terminus of Road 85. This gentle summit is covered with a large meadow that fills with wildflowers and butterflies in midsummer.

Farther south there's an area filled with colorful place names, the watershed of Ramsey's Draft. This tumbling, clear stream flows from Hardscrabble Knob, Tearjacket Knob, and Freezeland Flat down through mature groves of hemlock and mixed hardwood trees, eventually joining the Calfpasture River. A hiking trail follows the valley of Ramsey's Draft, crossing the stream on many occasions. Don't expect to keep your feet dry if you hike here, and if it's rained recently expect to wade in above your knees. Woodland wildflowers are abundant along Ramsey's Draft in spring, when nesting birds, such as the ovenbird, veery, and black-throated blue warbler, defend territories with their songs. The creekside trail connects with trails on the surrounding mountains. My most memorable wildlife sighting from this region was a great cluster of timber rattlesnakes I found around the rocky summit of Hardscrabble Knob on a midsummer hike.

Continue south and west from Ramsey's Draft to reach Bath County, Virginia. Place names of the county, such as Hot Springs, Warm Springs, Bath Alum, Healing Springs, and Blowing Springs, hint at the reason for the county's name. Massive sedimentary beds of rocks that were created 300 million to 600 million years ago underlie this section of the Ridge and Valley. The geological tumult of the building of the Appalachian Mountains 200 million to 300 million years ago caused much folding and cracking of these layers. Deeply fractured limestone beds occur beneath Bath County; water seeps down through these cracks and reaches depths where Earth's temperature is quite high. The heated water is pushed back up through the limestone, where calcium carbonate and other minerals are dissolved. When the water reaches the surface at a spring, it is rich in minerals and is often still warm. During the nineteenth century a number of resorts developed around these springs, whose waters were thought to have healing properties.

Bath County offers more than a warm mineral bath to the visiting naturalist. Lightly traveled backroads lead into the George Washington National Forest, past the forests, creeks, and small lakes of Douthat State Park, and through pastoral countryside where small farms and open pastures alternate with abandoned land where meadows and young forests grow. Bath County is one of the northernmost locations of a spectacular butterfly, the diana fritillary. July is the peak month for dianas, one of our largest and most beautiful butterflies. Watch for the orange and black males around flowering patches of milkweed, whose nectar they love. Female dianas are colored differently from the males; they are primarily black with a blue sheen to the hind wing. Females spend less time at flowers and more time cruising through open forests.

Bobcat

Where the Blue Ridge Parkway ends near the town of Waynesboro and not far west of Charlottesville, another mountain highway continues to lure the scenery-loving motorist northward. This is the Skyline Drive, which continues for just over 100 miles through Shenandoah National Park, a long, skinny park that hugs the Blue Ridge and its associated spur ridges.

Shenandoah differs from most national parks because, instead of an area whose primeval beauty had been protected, Virginia's Blue Ridge had been logged, farmed, and degraded environmentally for generations before the 1933 founding of the park. Those forests that had not been logged used to have many American chestnut trees, but by 1930 a fungal blight had killed almost every mature chestnut. Shenandoah was a restoration project. From its beginning there were fine views to enjoy, and the Skyline Drive was constructed along the ridge to be the central highlight of the park. Fields and pastures dominated the landscape; only 2 percent of the park was forested at its founding, and soil erosion had been severe in many places. But the eastern deciduous forest community is resilient, and the process of succession began. In limey fields the first trees were often red cedar; dry sites harbored Virginia pine and, in high places, table mountain pine. The tuliptree was a dominant species during the early stages of succession in moister lowlands. Shade-tolerant trees followed, with chestnut oak and red oak especially abundant.

The park today supports a forest that is gradually approaching maturity, though the trees are still relatively small in most places. Forest wildlife has returned, with white-tailed deer now extremely common, black bear populations seeming to increase, and bobcats sometimes seen. A few valleys that effectively trap more than their share of precipitation harbor southern hardwood cove forests, with diverse sets of trees, including hickories, maples, basswood, black gum, cherries, and many others. Dense stands of hemlock trees are found in other valleys, and small wetlands occur near some springs and streams, but most of the park is covered with the dry, oak-dominated forest.

The richer woodland soils bring the greatest wildflower displays, and in May the park dazzles. A favorite hike of mine leads east from the Appalachian Trail near the park's center and follows the Laurel and Mill Prongs to Camp Hoover, a fishing retreat of President Herbert Hoover in the days before Shenandoah was a park. At the peak of the May wildflower season this trail offers resplendent sights of yellow ladyslipper, trilliums, blue cohosh, and many others. Farther north lies a little side trail off the ridge crest to Gravelly Springs Gap, where the park's common wildflowers seem particularly prolific.

Waterfalls are a major park feature, as dozens of streams drop spectacularly off the Blue Ridge into the valleys below. The Appalachian Trail generally stays high on the ridge in Shenandoah, but beautiful falls are reached on side trails (all easily accessed from the Skyline Drive) throughout the park. White

Oak Canyon features six different major falls. Lewis Spring Falls, Rose River Falls, and the cascade known as Big Devil's Stairs are other popular destinations along the park trails.

Migrant songbirds find this north-south ridge with its continuous forest an attractive corridor, with May and September the peak songbird months. Catch one of the best days and the upland woods can be filled with scarlet tanagers and rose-breasted grosbeaks. Come nesting season there's a fine assortment of warblers and other songbirds on territory here. Chestnut-sided warblers are all over the semi-open ridgetops. Hooded, Kentucky, and worm-eating warblers love the densely forested slopes. Red-eyed vireos are everywhere, their insistent calls continuing all day long. Be careful to distinguish this call from the similar but higher and squeakier song of the solitary vireo, common in Shenandoah as a migrant but scarce during the nesting season.

North of Shenandoah National Park, the Appalachian Trail winds along a lower stretch of the Blue Ridge for about 50 miles, the last few straddling the West Virginia-Virginia boundary, before it reaches Harpers Ferry, crosses the Potomac, and enters Maryland. There's no scenic highway running alongside the trail in this section, but there are plenty of access points, as roads, small housing developments, and other development become widespread. Still, there are some outstanding natural areas to visit in this section, one of the best being G. Richard Thompson Wildlife Management Area, which is fewer than a dozen miles north of Shenandoah Park. Somehow this region manages to catch and hold more moisture than most areas of the Blue Ridge, perhaps because the slopes aren't as steep as on many other sections of the ridge. These gentler slopes are subject to less intensive erosion, and as a result the soil at Thompson is better developed and deeper than at most other Blue Ridge sites. This soil serves as a moisture reservoir, enabling a more diverse plant community to exist at Thompson than in many surrounding areas. Perhaps there is also some microclimatic influence. I've seen no data showing greater precipitation along this section of the Blue Ridge, but I have observed that the mountain here is often fog-shrouded.

Thompson's richness and diversity shine in early May, when one of the mid-Atlantic's most spectacular wildflower displays occurs. The Appalachian Trail runs through Thompson, but the area is also accessed by Virginia Route 638 north of Linden. After climbing steeply for a few miles past houses and pastures, the road turns to gravel and plunges into a fairly mature

deciduous forest. I'll never forget my first May
visit here. As soon as the road entered the woods,
I came screeching to a halt, jumped out of the car,
and just stood and marveled. The elegant wildflower called large-flowered trillium, which is
usually vivid white when fresh and later fades
to a soft pastel pink, was everywhere, literally
carpeting the ground. Some of the blossoms
were rich pink even though they were obviously fresh. I've long loved trilliums, and I'd
never seen such abundance before. I took
picture after picture before reluctantly deciding
to continue up the road. I was flabbergasted to discover that the carpet of trillium continued for miles—
there must be many millions here. When I walked on
I discovered marvel after marvel: yellow ladyslipper,
showy orchis, wild geranium, wood anemone, green violet (which I'd only seen in the Smoky Mountains previously), dozens of marvelous wildflowers in peak bloom, all
in abundance more spectacular than in any botanical garden. The wildflower display continues through the summer,
when Canada lily, showy skullcap, woodland sunflower, and

Yellow ladyslipper

many other species bloom. The richness of Thompson's wildflower show is
reflected in its wildlife, which is also varied and diverse. A search for reptiles
and amphibians at Thompson can yield views of two-lined salamanders and
spring salamanders in the creeks and five-lined skinks and eastern kingsnakes
in the upland woods. Nesting songbirds seem as common here as at Shenandoah National Park. As at many state-owned wildlife areas, hunting is permitted at Thompson, so check game regulations and seasons before ventur-
ing out in autumn.

Parts of the Appalachian Trail corridor in northern Virginia, Maryland,
and southern Pennsylvania were closed for a time in the 1970s. Large parcels
of land, where permission had been given for the trail to cross, were subdivided into smaller tracts and some landowners withdrew permission for the
trail to pass. Long sections, in one place about 20 miles, were rerouted onto
public roads for many years, until new agreements and land acquisitions
could be made to find a home in the woods for the trail again. Happily this

process was completed in the late 1980s, and the trail's right-of-way through the mid-Atlantic is secure.

During the 1970s, however, trail organizers weren't sure that the corridor would be preserved, so they began routing a long alternative trail along ridges farther west. In a worst-case scenario, the Appalachian Trail could be moved to this new route. This alternate, called the Tuscarora Trail in Pennsylvania and Maryland, and the Big Blue Trail in Virginia, now provides another long hiking route through the mid-Atlantic's Ridge and Valley region. A side trip along these trails takes us to more of the region's natural highlights.

The Big Blue Trail splits off from the Appalachian Trail in Shenandoah National Park, dropping quickly off the Blue Ridge and into the Shenandoah Valley. The valley is a long, broad expanse of low country that has long been a major agricultural area. Small towns and a few small cities are scattered throughout the valley, but pastoral scenery still dominates. Unfortunately there isn't much public parkland to explore. The best journey for naturalists through the Shenandoah Valley is a canoe trip down either the north or south branch of the Shenandoah River itself. Some stretches are quiet and placid, others feature a few rapids, but all offer good transects through the valley's mixture of farms, pastures, overgrown meadows, and forest patches. The river's aquatic ecosystem survives in spite of agricultural runoff and many abuses from the valley's industries; my fishing friends love certain sections, though they never tell me about the best fishing spots. On the banks of the river, as on the edges of the valley's quiet backroads, summer butterfly watching can be excellent. Much of the valley is underlain by limestone, with hundreds of caverns known to exist. Wild caves beckon the careful and trained explorer, and several commercially run caverns make the underground world accessible to everyone.

The Shenandoah Valley is divided into two sections by an interesting mountain range called the Massanutten. This range stretches for about 50 miles between the North Fork of the Shenandoah River, which lies to its west, and the South Fork of the Shenandoah, which lies to its east. Most is in public ownership as part of the George Washington National Forest. The Massanutten parallels the Blue Ridge, and was for a time the lead contender for the Shenandoah National Park, which eventually was created on the Blue Ridge instead. Although the Massanutten is never wide, in places it has four distinct parallel ridges in the chain. At both ends the range rises dramatically from the unbroken flatlands of the surrounding Shenandoah Valley; the southern terminus is Massanutten Peak, a beautifully shaped mountain when viewed from any angle and which looks a bit like Hawaii's famous Diamond Head Peak. At the northern end, where the range's two main ridges are bisected by Passage Creek and the Fort Valley, the western Massanutten ridge ends at the blocky summit of Signal Knob, and the eastern ridge at the angular outcrop known as Buzzard Rocks. Both summits are easily reached by well-maintained hiking trails.

The Big Blue Trail zips across the eastern Shenandoah Valley to reach the Massanutten Mountains as quickly as possible, entering the range at Veach Gap, where it turns right to follow Massanutten's eastern ridge north almost to Buzzard Rocks before it drops to cross the Fort Valley at Elizabeth Furnace. It then follows Little Passage Creek back to the south before dropping off the Massanutten and continuing west across the rest of the Shenandoah Valley. In its crossing of the Massanutten, the Big Blue Trail passes by several of the range's important biological communities. Like the neighboring Blue Ridge in Shenandoah National Park, the Massanutten is an area of relatively low rainfall, and dry woods dominate the ridges and slopes. Chestnut oak, red oak, and other hardwoods make up the typical forest, with some of the sandstone ridgetops covered by table mountain pine. This tree, which is restricted to the dry ridges of the mid-Atlantic, generally grows with a gnarled, twisted shape that rather resembles an oversized bonsai. Shales are found at the surface in many places, and several shale barrens, replete with rare shale barren endemic plants, occur within the Massanutten, one just off the road near Elizabeth Furnace.

Passage Creek, Little Passage Creek, and the other streams of the Massanutten follow rich little valleys through the range. In contrast to the dry forests on the surrounding ridges, these valleys support rich floodplain forests and a few hidden bogs where orchids and other uncommon wildflowers grow. The Massanutten is a range designed for wandering explorations. Take a day, a weekend, a week, or more and start following each small backroad and trail. There are delights to find around almost every corner.

The first prominent ridge west of the Shenandoah Valley is called Great North Mountain, and the Big Blue Trail climbs into this mountain not far from the high spot with the intriguing name of Big Schloss. Wild turkeys and white-tailed deer are extremely common here; use caution during the hunting season. West of the Shenandoah Valley, the Ridge and Valley region takes a different character. The ridges lie much closer together, and the intervening valleys are much narrower. The trace of the Big Blue Trail follows the ridges along Great North Mountain and Sleepy Creek Mountain, and crosses valleys, too, en route to Hancock, Maryland, where it becomes the Tuscarora Trail. From here the view west is dominated by the long, straight ridgeline of Sideling Hill. This mountain can't be mistaken for any other because a huge bite was taken out of Sideling Hill to allow for passage of Interstate 68. From a distance this road cut is astounding; it is immense, cutting the mountain almost in half. Up close, however, the Sideling Hill road cut offers a textbook geology lesson.

The Ridge and Valley is a region of folded sedimentary rocks. When I first learned this and looked at the map it seemed eminently sensible; the landscape looks folded with all these parallel ridges and valleys. Where the land folded up, I thought there would be a ridge, and where the land folded down, a valley. Sideling Hill shows, incredibly, that just the opposite is true. The bottoms of the old folds had their rock compressed, making them more resistant to erosion. The rocks on the tops of the folds were stretched, cracked, and weakened, and thus were more susceptible to erosion. Two hundred million years is a long time for mountains to wear down, and the differential rate of weathering has left us with ridges at the old fold bottoms (geologists use the term *syncline* for such a feature), with valleys at the former high spots. This type of landscape is called inverted topography. The state of Maryland operates a fine visitor center at the Sideling Hill road cut with displays that interpret the geology and general natural history of the Ridge and Valley region.

Several other sites in Maryland's Ridge and Valley are worthy of exploration. Just beyond Sideling Hill lies the Green Ridge State Forest, whose varied habitats are discussed in the Potomac chapter. Rocky Gap State Park is farther west, almost to Cumberland. At first glance this is simply a recreation park, with a reservoir for swimming and boating, a golf course, and a large annual country music festival. In the southwest corner of the park, however, is the Rocky Gap State Wild Land. *Wild Land* is Maryland's equivalent of federal wilderness, and state lands so designated are protected from development or resource exploitation. A remarkably narrow, steep gorge cuts through the ridge here, and several short trails give the visiting naturalist a chance to see stunning vistas, cliff-edge plants, and a surprisingly rich wildlife community. Bobcats still roam through Rocky Gap, though I've never been lucky enough to see one here myself.

As it heads northeastward through Pennsylvania, the Tuscarora Trail follows long, narrow ridges that are typical of the Ridge and Valley region. Along the summit of Tuscarora Mountain and then Blue Mountain, dry oak forest predominates, mixed with hickory, tuliptree, maple, cherry, and black locust. Much of the land is protected as state game land. All sections provide a fine woodland walk. Finally, after its nearly 250-mile detour, the Tuscarora-Big Blue Trail reconnects to the Appalachian Trail near Dean's Gap, close to the town of Carlisle. Because the original corridor of the Appalachian Trail was successfully protected, the Tuscarora-Big Blue route remains a lightly traveled alternative. Long hike or short stroll, this trail has much to offer the curious naturalist.

Chestnut-sided warbler

Before we ventured west on our long side trip, we had wound our way north on the Appalachian Trail to Harpers Ferry, where Maryland, Virginia, and West Virginia all come together at the confluence of the Shenandoah and Potomac Rivers. The Appalachian Trail leads right through the old town after dropping off the Blue Ridge just south of the Potomac at Loudoun Heights and crossing the Shenandoah. Side trails along the ridgetop lead to several spectacular vistas over the gorge. Chimney Rock juts up above the hillside forest directly across the Shenandoah from Harpers Ferry, the town's historic buildings seeming close enough to touch. Split Rock is a couple of miles farther east, offering views down the Potomac to Point of Rocks and across the broad Piedmont. The Appalachian Trail crosses into Maryland on top of an old railroad bridge over the Potomac, and near here yet another spectacular side trail winds upward to the overlook called Maryland Heights, where it feels like you're floating right above the town. For many years this has been one of my favorite retreats; it's a great place to watch the sunset in June, when the mountain laurel is in bloom and the nesting chestnut-sided warblers and Acadian flycatchers are singing. Swallows, chimney swifts, and common nighthawks swoop over the river far below. Beyond the Great Valley to the west, I can watch the sun disappear behind the outline of several distant ridges. All of the Harpers Ferry over-looks can be reached with modest hikes of just a few miles each, and several longer loop hikes can easily fill a weekend.

For a little over a mile the Appalachian Trail follows the C & O Canal towpath, then it climbs up to the summit of South Mountain, which it fol-lows for almost 40 miles to the Pennsylvania state line. Elevations are modest on South Mountain, and those who love lofty peaks sometimes find this stretch of the Appalachian Trail a disappointment. Even though South Mountain's summit is only around 1,000 feet (300 meters) in elevation, Maryland's section of the trail offers a long, fairly level ridgetop walk through an open oak forest. A few outcrops of the greenstone rock of South Mountain offer expansive views west across the Hagerstown Valley. Where the vistas point east, Catoctin Mountain comes into view. A number of small housing clusters occur along the Catoctin, not surprising given its proximity to Frederick, Washington, D.C., and Baltimore, yet much of the land is pro-tected. The best-known mountaintop hideaway on the Catoctin is one I've never visited, the presidential retreat of Camp David, but nearby are the fed-erally protected Catoctin Mountain Park, state parks at Cunningham Falls and Gambrill, and a large region on the mountain's eastern slope that is pro-tected as forest for the city of Frederick's public water supply. Little back-roads and trails through the Catoctin are fun to explore. Although the forest is dry and oak-dominated like most forests in the Ridge and Valley region, several creek valleys seem to maintain moisture and are excellent sites for wildflowers, birds, mushrooms, and, if you turn over a few rocks in the creeks and along their banks, salamanders. Trout thrive in these creeks; my grandfather taught me how to fly-fish in the Catoctin Mountains. The trail along Owens Creek in Catoctin Mountain Park is a favorite, and at times of high water or freezing weather Cunningham Falls is surely worth a visit.

Back on South Mountain, the Appalachian Trail crosses two state parks, which can also be accessed by road. Gathland State Park sits in Crampton's Gap, a notch along the ridgetop about 7 miles north of the Potomac. This park preserves a stone arch built in 1896 by George Alfred Townsend as a monument to war correspondents. Townsend was an internationally famous writer whose career began during the Civil War and continued into the twentieth century; he often wrote under the pen name of Gath. He built an extensive estate at the gap in addition to the memorial arch. After Town-send's death in 1914 most of the estate buildings fell into disrepair. By the time Maryland acquired the land and created Gathland State Park in 1958, all that remained were the arch, a few buildings and the ruins of others, and an empty mausoleum with the words, "Good night, Gath," inscribed over the door. Townsend built the mausoleum for himself, but he left his estate late in life to live with his daughter in New York, where he died and was buried.

Washington Monument State Park, about 10 miles north of Gathland, may be accessed by road from Alternate U.S. Route 40 or on foot by the Appalachian Trail. On the summit of Monument Knob, the park's highest point, is an odd stone tower, roughly tubular in shape and about 30 feet high. The tower was originally built in 1827 as the nation's first monument to George Washington. The tower has collapsed and been rebuilt several times, the last thorough reconstruction done by the Civilian Conservation Corps in 1936. Climb the spiral stone stairs for a view west to Boonsboro, whose Whig party activists built the first tower here. Visit in autumn, however, and you'll surely find a few people staring intently at the sky with binoculars. Washington Monument is a great hawk-watching spot, and from September through November one or more knowledgeable volunteers are stationed at the monument virtually every day, keeping a tally of all southbound raptors that are seen.

Most of our eastern hawks and related birds of prey are at least somewhat migratory. They are highly energy efficient, migrating primarily where they can travel the greatest distance with minimal effort. The linear, roughly north-south oriented ridges of the mid-Atlantic deflect the prevailing westerly winds upward, especially after strong autumn cold fronts pass through. Migrating hawks find these updrafts and stay with them, gaining altitude effortlessly by just gliding along. Sometimes they slip off the updraft and are blown eastward. In most of the Ridge and Valley this is no problem because the birds just take the tailwind over to the next ridge east. But eventually the birds run out of ridges and end up doubling back to the last continuous ridge. So although each mid-Atlantic ridge becomes a migratory thoroughfare, the easternmost ridge collects far more hawks than the others, and it's along this ridge that most of our region's premier hawk-watching sites are found. In Maryland it's South Mountain, with Washington Monument the favored viewing spot. In Virginia the Blue Ridge becomes the primary flyway, with two spots along the Appalachian Trail—Snickers Gap and Rockfish Gap—especially well-suited for viewing the migration. An excellent hawk-watching spot in Pennsylvania is Waggoner's Gap, which is along Blue

Mountain just off the Tuscarora Trail near Carlisle. Farther north and east, just a stone's throw from the Appalachian Trail, is the granddaddy of all hawk-watching sites—Hawk Mountain on Kittatinny Ridge, some 20 miles north of Reading.

Even though hawks can be seen migrating anywhere over the eastern ridges, viewing is best where the mountain dips, the updrafts of wind weaken, and the hawks necessarily drop to lower altitudes, where they are more readily observed. These are the conditions found at Hawk Mountain, which also features open rocky mountaintop observation sites that provide unobstructed viewing of the oncoming hawks. In the early twentieth century all raptors were considered vermin, and gunners would gather at Hawk Mountain to shoot thousands of hawks, falcons, and eagles. Attitudes began to change in the 1920s, and in the 1930s a group of conservationists purchased the land at Hawk Mountain and created a private nature reserve. Raptor shooting was eventually outlawed and hawk populations began to increase, monitored by the staff and volunteers at Hawk Mountain and other sites. As the site's reputation grew, more people would visit each year. Annual visitation is now many thousands of persons, some avid hawk watchers who can identify seeming pinpoints in the sky, others just curious passersby stopping to see what all the excitement is about. In addition to the birds, Hawk Mountain now offers a nature center filled with informative displays and an excellent series of education programs year-round. It also supports a variety of research projects on raptors and other birds on the sanctuary and across the globe. I learned to identify raptors at Hawk Mountain, thanks largely to the staff naturalists who were always eager to help, and was lucky enough to be there for a record day, when around 2,800 sharp-shinned hawks were spotted. I don't get to Hawk Mountain as often as I used to, but I can't watch an autumn cold front pass by without yearning to go.

So ends our tour of the mid-Atlantic's Ridge and Valley region. The Appalachian Trail provides a thread to link many of these sites together; it links them to natural areas in Connecticut, Vermont, and right up to Maine. The adventurous naturalist can become intimately acquainted with the Ridge and Valley through a long hike, but many places on and near the Appalachian Trail are also accessible to nonhikers. By whatever means you choose to explore the Ridge and Valley region, take time to look beyond the great mountain vistas and see how these drier mountains differ from those found to the west and from the lowlands to the east.

9

The Bay—
Exploring Lands around
the Chesapeake

In Seattle, when people talk about Mt. Rainier they simply refer to *the mountain*. There's no mountain in the United States (outside of Alaska) that matches the grandeur of Rainier, which looms high above Seattle. Talk about *the city* anywhere in the Northeast is surely a reference to New York. Love it or hate it, few would deny that New York City is a unique place, the ultimate city. Here in the mid-Atlantic we talk about *the bay*. When we do, everybody knows we're talking about that grandest of all estuaries, the Chesapeake Bay.

Delmarva fox squirrel
in loblolly pine

When it comes to a discussion of the bay, I'm at something of a loss, for I've never owned a boat. Although I've been out on the waters of the bay and its tributaries in a variety of craft, enjoying a peek at the underwater life by dragging a plankton net or a fishing lure through the water, my personal relationship with the Chesapeake is that of a terrestrial explorer. Thankfully there are many natural areas where the bay's shore, its extensive wetlands, and surrounding upland habitats may be explored.

The Chesapeake Bay is a textbook estuary, a semi-enclosed body of water where fresh water from rivers mixes with salt water pushed in from the ocean by the tides. The bay stretches for 200 miles (320 kilometers) from the mouth of the Susquehanna River near Havre de Grace, Maryland, at the north, to its junction with the Atlantic near Norfolk, Virginia, at the south. Its average depth is just 21 feet (6.4 meters), its width varies from 4 to 30 miles (6–50 kilometers), and its waters cover about 2,500 square miles (6,500 square kilometers). The bay's convoluted shoreline extends for about 4,000 miles (6,400 kilometers), with about 500,000 acres (200,000 hectares) of wetlands lining its shores and extending up its tributaries. Its watershed includes the drainages of the Potomac, James, York, and Rappahannock Rivers in addition to the Susquehanna, an area of 64,000 square miles (166,000 square kilometers).

Geographically the bay follows the ancestral floodplain of the Susquehanna River. During ice age periods, when so much of the Earth's water was tied up in the massive polar ice caps that sea level was hundreds of feet lower, the region covered by today's Chesapeake was well above sea level, as was much of the mid-Atlantic continental shelf. The Susquehanna cut a two-part valley through the region, as most rivers do. The lowest part of the valley was the river channel itself, with the river's broad floodplain perched above the channel but still below the surrounding countryside. Every river and creek that fed into the Susquehanna similarly carved a floodplain in addition to a channel. The entire Chesapeake, like all the land to its east and some to its west, lies within the Coastal Plain geographic province, a region where bedrock lies well below the surface, buried under sediments that have eroded off the Appalachian Highlands for 200 million years. Rivers and creeks have had little trouble carving valleys through these sediments.

The most recent ice age was winding down about 12,000 years ago. The melting of most of the polar ice caps led to the gradual rise of the oceans. The sea stabilized about 3,000 years ago, at a level roughly 325 feet above its ice age low. The last 200 miles of the Susquehanna floodplain was inundated by the ocean, along with sections of most of its tributaries, and these flooded rivers form today's Chesapeake Bay. Because the opening to the ocean is fairly small compared with the bay's length and volume, the flow of salt water into the bay is somewhat restricted. Combine the constant flow of fresh water into the bay from the Susquehanna and the other rivers with the restricted inward flow of salt water and the result is a large body of brackish water. The bay is salty, though everywhere less so than the ocean, and the salinity varies dramatically from north to south and from season to season. It's amazing how many marine organisms have evolved to spend part of their lives in the unique brackish-water environment of estuaries like the Chesapeake.

The flooded tributaries, and in places the edges of the bay itself, are lined with vast areas of wetlands. Brackish marsh is most widespread in tidal areas, but there are salt marshes near the bay's mouth and freshwater marshes and swamps on many of the tributaries. Wetlands are among the most vibrant and diverse ecosystems in the bay region—from a distance they can look as ecologically boring as an Illinois cornfield, but up close you can see they are just exploding with life. Ecologists compare environments with measures of their productivity, defined as the amount of the Sun's energy reaching Earth at each spot, which is transformed into chemical energy by green plants and then assimilated into food chains. Tidal marshes, both salt and brackish, are among the most productive ecosystems on the planet. The marsh grasses are not all the same—in fact not all are true grasses, but are an intriguing mix of big cordgrass, saltmarsh cordgrass, saltmeadow hay, switchgrass, needle rush, olney three-square, and a few others. The lowest places in a tidal marsh are covered with water most of the time, whereas the highest are covered by just the highest tides. Some marsh plants are best suited to the low spots, others to the high, and still others to the middle zones. The minute variations of elevation that occur in each marsh thus result in distinct zones where different plants dominate. Marsh-dwelling animals that live in the water, above the water, or in the wet mucky soil generally move from one zone to another with the changing of the tides and of the seasons. Many also use surrounding aquatic and terrestrial habitats. Because each zone provides different sources of food and shelter to animals, a great variety of life is able to thrive in these tidal marshes. Birders visiting Chesapeake wetlands know to look for saltmarsh sharp-tailed sparrows, marsh wrens, shorebirds, herons, egrets, and bitterns during the day, and to listen for the calls of rails in the bay's marshes

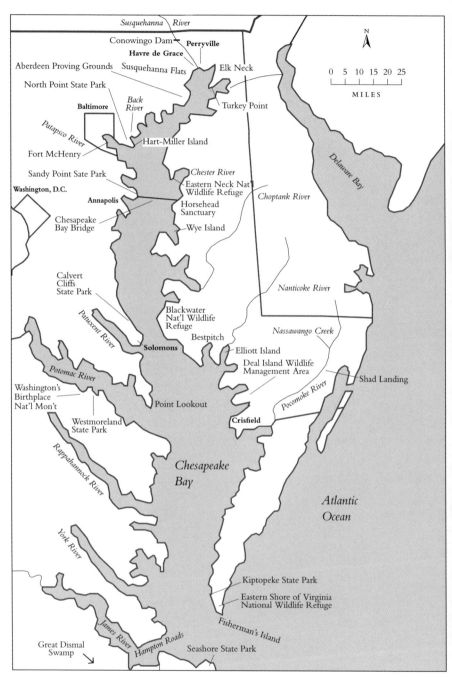

Map 5.   Chesapeake Bay Region

at night. But the greatest diversity of the marsh occurs underwater, where abundant animals live, ranging from big fish, crabs, and mussels to the microscopic plankton.

The uplands that surround the bay and its tributaries, although not "up" very far, support a rich forest ecosystem that in many areas has been replaced by rich agricultural fields. Earlier in the twentieth century the farms of the Eastern Shore, the regional name for those lands lying east of the bay, supplied table vegetables for the truck and canning trade up and down the East Coast. This industry faded with improvements in transportation systems and the development of truck farming in warmer parts of the country such as Florida and California. Today these fields are mostly planted in corn, wheat, and soybeans. Soybeans are used primarily as feed for chickens, a major segment of the Eastern Shore's agricultural industry.

Many patches of forest have been preserved, and efforts to protect the health of the Chesapeake have included incentives for landowners to keep major areas within the bay's watershed in forest, especially sections adjacent to tributary creeks. The idea is to reduce the runoff of sediment and agricultural chemicals into the bay. Some wonder if this and other efforts to protect the bay may be too little and too late; the second half of the twentieth century has seen a dramatic decline in many of the bay's organisms, including oysters, submerged aquatic plants, shad and other finfish, and even the pollution-tolerant and seemingly resilient blue crab. The bay is still a scenic treasure, and it still harbors a lot of life, but its health continues to decline. Given its importance in the region's economy, history, and culture, efforts to protect the bay theoretically receive widespread support, but in practice we've got a long way to go if we're to preserve much of the bay's richness.

The headwaters of the Susquehanna River, source of about half the bay's fresh water, lie in southern New York state, passing near Binghamton and then through a long stretch of Pennsylvania's Ridge and Valley region. The river winds past industrial towns and through coal-mining country before reaching the Piedmont near Harrisburg, Pennsylvania's capital. Just downriver lies the imposing structure of the Three Mile Island nuclear power plant, where a crisis in 1979 nearly resulted in disaster. From here to the bay the Susquehanna, now a sizable river, winds through a rolling mix of farm country and forested blocks, passing right between the small cities of York and Lancaster. Four old hydroelectric dams along the river's lower stretch enlarge its current channel and, sadly, block the movements of many fish that formerly moved between the river and the bay. The antiquated Conowingo Dam, near Maryland's northeastern corner, has become a great spot for birding for a rather sad reason. Many fish are killed by the power-generating turbines of the dam, and in the colder months bald eagles, common mergansers, and thousands of gulls, both common and rare species, congregate below the spillway to feed on the shredded fish. The power companies that operate the Susquehanna's dams are trying to correct these problems. Shad, once an extremely abundant fish of the Chesapeake Bay, and which travel from salt water to fresh in order to spawn, are now physically moved around Conowingo Dam, and their populations along the Susquehanna are increasing.

It's only about a dozen miles along the river from Conowingo to the bay. Susquehanna State Park protects riverfront and upland forest habitat along this stretch. Great spreading silver maple trees line the river near an old stone mill, where the park staff offers historical demonstrations. A little butterfly called the meadow fritillary seems especially fond of the park's open fields and picnic areas in early summer. The towns of Havre de Grace and Perryville lie across the river's mouth from one another. Views south from both towns open onto the vast, shallow expanse of the upper bay, a region called the Susquehanna Flats. This section of the bay is protected as the Susquehanna National Wildlife Refuge, because historically the flats were filled with submerged aquatic vegetation, particularly a plant called wildcelery. In fall and winter massive flocks of waterfowl gathered here. Historically the Susquehanna Flats was the winter home to huge flocks of canvasback, the elegant duck most often used as a symbol for the bay's waterfowl legacy. In December 1952, 91,000 canvasbacks were counted here. Now, pollution in the bay has resulted in a devastating decline in its aquatic plants, with increased turbidity of the water a major problem. Sediments are carried into the bay from construction sites, farms, and other places where vegetative cover is removed. Nutrients flow in from farms, suburban yards, and with the added effluent from sewage plants, the overnourished floating algae of the bay's waters then overgrow. Sediments and algae both block the passage of light through the water, and it seems that in the Susquehanna Flats, as around most of the bay, so little light penetrates that few, if any, submerged aquatic plants can survive. Peak season canvasback counts for the entire bay are now below 70,000; usually there aren't any canvasback at the Susquehanna Flats.

East of the Susquehanna Flats is a tapered peninsula called Elk Neck. Some of this land is privately owned, but a state park and a state forest also occur here, both called Elk Neck. Each offers access to some varied Coastal Plain forest. The tip of the peninsula is called Turkey Point, and is part of the park. A short trail leads to an old lighthouse near the land's end, a high bluff about 100 feet above the bay. Views down the bay on a clear day seem endless. As at other southward-pointing peninsulas, migrant birds concentrate over Turkey Point during autumn, with raptors often the most conspicuous. A recent banding study revealed that northern saw-whet owls also migrate over Elk Neck in significant numbers. Both migrant and resident songbirds are plentiful here, with eastern bluebirds happily quite common.

Heading down the bay from the Susquehanna Flats, the western shore is dominated by the Aberdeen Proving Ground, a large military installation. Access to the lands at Aberdeen has long been restricted, and as a result much of the area's forest and wetland habitats are fairly undisturbed. Bald eagles don't tolerate much human activity, and some of the region's highest counts of both nesting and wintering eagles are from Aberdeen, thanks no doubt to its isolation. The eastern shore of the upper bay is mostly rural, though suburban development is slowly spreading west from Wilmington into a few communities along the Sassafras and Elk Rivers.

Southwest of the Aberdeen Proving Ground, the bay takes on a very different character: Baltimore. The bay's biggest city has plenty of charm, but by definition an urban area has major effects on its surrounding environment. Sparrows Point, southeast of the city, is home to a huge steel mill, historically the source of heavy metals and other pollutants into the bay. Just east of Sparrows Point, however, is the unspoiled North Point State Park, which protects Black Marsh, a large tidal wetland. Debates have been raging about the future of the park, whether it will be maintained primarily as a natural area or developed for recreational purposes. I'm hoping for some sort of compromise that will leave the marsh relatively undisturbed. Just minutes from the Baltimore Beltway, this park offers plenty of opportunities for city dwellers to see the vibrancy of a brackish marsh and to scan the open waters of the bay. Within view from the shoreline here is Hart-Miller Island, accessible only by boat. Although Hart-Miller is a repository for much dredge spoil from Baltimore Harbor, it has become a major stopping point for migratory shorebirds—an incredible variety are seen here each year. Birds that have tried nesting here have had little success, however. Some speculate that nest failures are caused by accumulated toxins from the harbor bottom that are part of the dredge spoil material, but no one knows for sure.

Sparrows Point and North Point are part of a peninsula separating the Patapsco and Back Rivers. The Back River's short tidal stretch is lined with a lot of development, including a large sewage treatment plant and an old landfill. Curiously the sewage plant has become a birdwatcher's haven, as gulls, crows, and other scavenging birds find nourishment in the tanks of partially treated sewage. In the colder months thousands of gulls are found here, most either ring-billed or herring, our region's two most common gulls. Occasionally, however, birders find a surprise—most noteworthy was the appearance one year of a Ross' gull, a beautiful bird that rarely strays from the Arctic Ocean. Workers at the plant were amused and delighted as thousands of eager birders visited Back River while the Ross' gull was present; eventually, these birders were joined by newspaper and television reporters drawn to the phenomenon.

The Patapsco River's tidal stretch is Baltimore's harbor, once quite grimy and polluted. A lot of progress has occurred in reducing point source pollution from the bay in recent decades. The cleaner water and the attractive urban development around the inner harbor make this a pleasant spot to visit, though it is still mostly a biological desert. The most interesting natural area along the Baltimore Harbor is Fort McHenry, the old fort of "The Star Spangled Banner" fame, now protected as a unit of the national park system. For several consecutive winters in the early 1980s a snowy owl wintered at Fort McHenry, and it's still a decent spot to watch migrating hawks. Fort Smallwood, down by the junction of the Patapsco and the bay, is an even better hawk-watching spot. Upriver from Baltimore, the Patapsco winds through a forested valley, most of which is protected as state park (see the Close to Home chapter for more details about Patapsco Valley State Park).

Heading south from Baltimore, the bay's best-known artificial landmark looms high: the Chesapeake Bay Bridge, which carries U.S. Routes 50 and 301. By far the easiest link from the cities of Baltimore and Washington, D.C. to the Atlantic Coast, the Chesapeake Bay Bridge and Route 50 have an enormous amount of traffic, especially in summer when visits to the beach are popular. Because the bridge rises high above the bay's main shipping channel, the view from the high point of the road is terrific, though the view is best to enjoy when somebody else is driving. Peregrine falcons nest on the cement pillars that anchor the central suspension section of the westbound span. Just north of the bridge are two interesting natural areas, one on each shore, Sandy Point State Park and the Eastern Neck National Wildlife Refuge.

Sandy Point, on the western shore, mixes recreational activities with the protection of some unspoiled natural habitat. A marina is popular with those who enjoy the bay for pleasure boating and fishing, and a tan sandy beach invites people to swim, windsurf, build sand castles, and play volleyball. Back from the bay's shore is a fairly large area that's seldom visited, a mix of woods, fields, and wetlands where a surprising variety of wildlife may be found. The bay itself comes alive in fall and winter when seafaring ducks such as the common goldeneye, oldsquaw, and bufflehead are often common. Watch also for big flocks of scaup, common loon, and horned grebe. Once in a while northern gannets, which generally stick to the ocean, may be seen here, and eared grebes, rare in our region, have been seen on the bay here several times. Canada geese can be counted by the hundreds around Sandy Point during the cooler months, when tundra swans are also fairly common. At any season it's worth stopping at Sandy Point at least briefly to watch for peregrine falcons and other raptors and to check through the gull flocks— some astounding rarities have been found here at all seasons. A small flock of snow buntings usually winters at Sandy Point. These black, white, and beige sparrows nest in the high Arctic of Canada and Alaska.

The Chesapeake Bay Bridge's eastern anchor is on Kent Island; just north of here lies the mouth of the Chester River, with Eastern Neck Island hugging its northern shore. The island is protected as the Eastern Neck National Wildlife Refuge. Although Eastern Neck is just a couple of miles from Kent Island and the Chesapeake Bay Bridge, it's an hour's drive away because no bridges cross the Chester downriver from Chestertown. Even though the route is circuitous, it's also pleasant, leading through quiet farms and towns, past marshes and forests, and finally winding down to the refuge. Eastern Neck is never crowded with people, though tundra swans and other waterfowl become abundant here in November. Earlier in the fall this is a great spot for watching southbound migrants, with many hawks, geese, and a variety of songbirds often passing through. Tree swallows sometimes stream south by the thousands. Red foxes seem especially common in this section of Kent County, and you can't watch a brackish marsh for long without seeing a muskrat or two. When my life has been especially hectic, and I'm looking for a spot where I can slow down and relax, Eastern Neck always comes to mind as an ideal destination.

Two other interesting natural history destinations are on the Eastern Shore near Route 50 and the Chesapeake Bay Bridge: Horsehead Sanctuary and

Wye Island. Horsehead, just across Kent Narrows from Kent Island and near the town of Graysonville, is owned by the Wildfowl Trust of North America, a private, nonprofit organization dedicated to education, research, and conservation of waterfowl. Horsehead has a visitor center with informative displays, along with an intriguing mix of captive waterfowl and natural habitat. Ponds near the visitor center are home to a variety of captive ducks and geese, but the ponds are attractive enough to entice many wild ducks to visit, too. Other large ponds are managed solely for the wild birds, and well-constructed observation blinds often allow excellent viewing of wild ducks and geese. The sanctuary also protects a large brackish marsh bordering the bay, a spot especially scenic in the late afternoon when the sun begins to set. It's a great place to watch bald eagles, red-tailed hawks, northern harriers, or ospreys circling overhead.

Wye Island is a sleepy little secret on the bay's edge a little farther south and east. A major housing and recreation development was once proposed for Wye Island, but thankfully, the plan never materialized and today the land is owned by the state of Maryland and protected as a natural resource area. Follow Carmichael and Wye Island Roads south from Route 50 to reach Wye Island. You won't find much on the island besides an old farm and a mixture of old fields, wetlands, and mature hardwood forest. It's a great spot for poking around on foot, looking for birds or butterflies, box turtles or blue crabs, but an even better way to see Wye Island is by canoe. It's a wonderful, full-day paddle to circumnavigate the island.

Much farther south, beyond the broad waters of the Choptank River, are the vast brackish marshes of Dorchester County, many of which are protected as the Blackwater National Wildlife Refuge. Blackwater has long been considered one of our region's top natural history destinations, its best-known attractions bald eagles and great flocks of waterfowl. The waterfowl numbers are down considerably at Blackwater as throughout the region—recent years have not been good for these birds—yet a November visit to Blackwater is still an essential part of many naturalists' calendars. Fields around the visitor center are often filled with big flocks of Canada geese and snow geese, many of the latter in the dark or *blue goose* color form. Bald eagles often perch on one of the snags behind the visitor center and cruise overhead. Experienced birders almost always tally twenty or more bald eagles on a fall or winter visit to Blackwater, and many years the refuge also harbors a wintering golden eagle or two.

I'll never forget my first visit to Blackwater National Wildlife Refuge, during my freshman year at the University of Maryland. My forestry professor brought the class here on a field trip, nominally to examine the Coastal Plain forest and discuss the refuge's forest management plan. It was mid-November, however, and Dr. Kuntz knew well that we'd be seeing the peak populations of geese. This was back in the mid-1970s, when Canada goose populations at Blackwater reached their historical highs, with nearly 100,000 birds present. We pulled into the visitor center parking lot and were immediately mesmerized; the cornfields were so full of geese that you could barely see the ground. I couldn't believe it was real until we stepped outside and heard the raucous cacophony of thousands of simultaneous honks. I don't remember much about the forests we observed that day, but I definitely remember the geese!

Blackwater's public access area consists of a fairly short wildlife drive and a few walking trails. The vast tidal marshes of the Blackwater are the obvious attraction, and there are always lots of birds out here. During the warmer months watch for herons, egrets, terns, and shorebirds. During colder months ducks, geese, swans, and gulls are more prevalent. Many raptors are always present, and great blue herons spend the entire year here. Blackwater's marshes are disappearing, however, and scientists are struggling to determine precisely why. Sea level continues to rise relative to the land level around the Chesapeake, which results in the inundation of marshes and low-lying islands throughout the bay. One theory suggests that the land is dropping more significantly than the sea is rising, perhaps partly due to the intensive agriculture of the Eastern Shore. Huge quantities of water are pumped out of the ground for agriculture, particularly to support the poultry industry. This theory suggests that the land was supported in small part by the aquifer lying underneath, and as the water has been systematically removed, the land has subsided. It may be just a few inches of subsidence, but in a marsh habitat a few inches can make a big difference.

Views across the Blackwater marsh now show more open water than when I first visited the refuge two decades ago, but there's still a lot of marsh remaining, thank goodness. Muskrats and the much larger, introduced aquatic mammal called nutria can often be seen swimming through the marsh, and on several occasions I've seen gray foxes at the edges of Blackwater's wetlands. Although the wetlands are the refuge's most obvious feature, significant blocks of forest are also protected here. The classic Coastal Plain forest types of the Eastern Shore are well-represented at Blackwater, with loblolly pine and a variety of hardwoods vying for dominance from place to place. In some areas the pines nearly create a monoculture in the canopy; these are places to search for brown-headed nuthatches. In other areas hardwoods are dominant, with white oak and sweetgum often the most common trees. Areas where the pines and the hardwoods mix are the favored habitat for one of our region's rarest mammals, the endangered Delmarva fox squirrel. A substantial population of these big, silvery squirrels remains at Blackwater. These squirrels are bigger and lighter than the ubiquitous eastern gray squirrel, with a flatter face and an extra long, bushy silver tail. They tend to spend more time on the ground than do gray squirrels, though they are also agile up in the trees. In the warmer months Blackwater is an excellent area for butterfly study, as a great variety visit flowers in open fields and wetlands. The aptly named rare skipper occurs around Blackwater, along with many more widespread species.

Most explorations of Dorchester County start and end at Blackwater. This corner of the Chesapeake has sleepy towns, quiet backroads, and expansive marshes to explore. Pick up a map and choose your corner. Whether you head through Bishops Head to Crocheron, down Hooper Island to Hoopersville, or out to Hudson, James, and Thomas near the mouth of the Choptank, you will find out why the Chesapeake country has long been called *the land of pleasant living.* My favorite adjunct to a Blackwater trip heads first to Bestpitch, east of the refuge at the crossing of the Transquaking River. I once took a long canoe trip down the Transquaking, falling in love with its great, untouched marshes dominated by big cordgrass. Once the weather turns cold, rough-legged hawks take up residence for the winter at Bestpitch.

From Bestpitch I head east over a stretch of road that's flooded a few times each month by the extreme high tides and head to Henrys Crossroads, where I turn south and head to Elliott. Marshes along the road to Elliott are richly varied, with different plants dominating in different stretches. In places the marsh seems endless. Much of the marsh is owned by the state and protected as the Fishing Bay Wildlife Management Area.

Birders are drawn to these marshes in late spring and early summer for one major reason, to search for black rails. I've made the quest several times, and it's like nothing else in the universe of nature study. A little background is perhaps in order. Rails are wetland birds that spend much of their time walking quietly through the marsh grasses. They tend to be much shorter than the grasses they are walking through, which makes them quite hard to see. Even though I know a couple of spots where king rail and clapper rail tend to behave rather conspicuously, in general I feel lucky any time I see a rail.

Gray fox

vck 76

Black rails are especially elusive for several reasons. They are very small, about the size of a sparrow. As their name implies, they are also dark, though their black background is decorated with white belly stripes, a rusty brown patch in the upper back, and a few white speckles on the lower back and wings. Also, these birds are primarily nocturnal and are rare. Now doesn't this sound like fun—going into a marsh at night to look for a black bird that stays hidden among the marsh plants! It would be impossible were it not for one thing: Black rails are vocal and, in late spring and early summer, fiercely territorial. Visit a black rail marsh in season, at night, and with luck you will hear a ringing *kee-kee-kurr* call. Call back at the bird (ah, the wonders of modern technology, such as a tape recorder) and, every once in a while, a territorial male black rail may come stalking out of the marsh to chase away the intruder—you. I have watched a little black rail vigorously flinging itself at my tape recorder; a friend thought he heard it say, "Get out of that box and fight like a rail!"

The road to Elliott offers much more than black rails. It's the best place I know in the mid-Atlantic for hearing, and occasionally seeing, whip-poor-wills and chuck-will's-widows. Also, the chorus of frogs in late spring in this wetland is delightfully varied and loud. There are wonders to explore during the daylight hours too. Bald eagles and other raptors can be seen over the marsh at any season, and in spring and summer the wooded patches around the marsh are home to many songbirds, including summer tanagers and blue grosbeaks, while the marsh itself is busy with the nesting activities of marsh wrens, seaside sparrows, and saltmarsh sharp-tailed sparrows. I especially love visits to this vast marsh on winter days, when sometimes the lightest dusting of snow is sprinkled over the floppy brown stalks of last year's marsh grasses and the cold breezes howl. Winter sunsets at Elliott can be spectacular, especially when short-eared owls fly low over the marsh with the brilliant orange sky behind them. Be sure to stop in at the little general store down in the town of Elliott for a snack, a drink, or some animated conversation, and visit the town's boat landing to enjoy the view across the bay's upper Tangier Sound. Watch for common loons and tundra swans from November through March.

Across the bay from Dorchester County lies Calvert County, at one time just as rural as Dorchester but now growing rapidly as the suburbs continue to sprawl out from Washington, D.C. and Baltimore. There are still some excellent natural areas along Calvert's bay shore, with Calvert Cliffs State Park perhaps the best known. A 2-mile trail leads through mature oak, beech, and pine forest to the edge of the bay, where impressive brown cliffs rise 120 feet (36.6 meters) up from a narrow beach. Clay layers along the cliff were laid down 12 to 17 million years ago, when all of southern Maryland was under a shallow sea. Fossils representing more than 600 species of marine organisms have been found here, including fossils of crocodiles, turtles, whales, and seabirds. Shark's teeth, oysters, and clams are especially abundant fossils. At Calvert County's southern tip is Solomons Island, located at the mouth of the Patuxent River. Solomons Island is a popular boating area because the bay waters near here are great for fishing. Birding is excellent around Solomons too, especially during the colder months.

The Patuxent River rises in north-central Maryland and flows halfway between Baltimore and Washington, D.C. on its way to the bay at Solomons. The Patuxent's watershed is heavily developed, though the river's floodplain corridor is protected as park or refuge in many places. Runoff from roads, houses, farms, shopping centers, and industrial areas combine with sewage effluent, making the Patuxent highly vulnerable to pollution. Many see the Patuxent as a model for the entire bay; if we can find a way for metropolitan development and a healthy Patuxent River ecosystem to coexist, perhaps a solution to the greater environmental problems of the bay can be found. For more on the Patuxent River parks, see the Close to Home chapter.

Continuing southward along the bay's western shore we come next to peaceful St. Mary's County and the mouth of the mighty Potomac River, which meets the bay at Point Lookout (described in the Potomac chapter). Virginia claims that section of the bay south of the Potomac's mouth. The Virginia shoreline consists of a set of three peninsulas divided by the broad tidal rivers that flow into the lower bay. South of the Potomac lies the Rappahannock; the land between these two rivers is known locally as the *Northern Neck*. This is lightly developed rural country steeped in history, with many old plantations where important colonial Americans were raised. George Washington's birthplace, now a national monument, was along Pope's Creek close to the Potomac. Stratford Hall, where Robert E. Lee was born, is nearby. Washington's birthplace and the nearby Westmoreland State Park are fine places to enjoy wildflowers in spring, butterflies in summer, and waterbirds in autumn and winter. The tidal Rappahannock has extensive wetlands to explore and is a great place for canoe trips. After the Rappahannock is the York River, with Yorktown on its south bank. Here the British surrendered and ended the Revolutionary War at a site now preserved by the National Park Service. Like so many parks across Virginia that are based on colonial or Civil War-era history, Yorktown also protects the natural environment and is a fine area for naturalists to visit.

Between the York and James Rivers lies a peninsula dominated by the cities of Newport News and Hampton, and across the James are the cities of Norfolk, Portsmouth, and Virginia Beach. The whole region has long been known as Hampton Roads, though it's the waterways that historically were the *roads*. Even though this is a major urban center, there are some fine natural areas lining this southern section of the bay. A variety of parks, refuges, and historical properties provide access to the James River between Richmond and Hampton Roads. Bald eagles are fairly common on the James, and peregrine falcons nest in a couple of spots. One year a fork-tailed flycatcher, a songbird of the American tropics, showed up in farm fields next to the James.

My favorite spot in the Hampton Roads area is Seashore State Park, which is close to the bay's mouth on the south side, and within the limits of the city of Virginia Beach. A thin beach and small, sparsely vegetated dunes line the park's northern boundary, the Chesapeake shoreline. Gaze out over the vast expanse of the bay here and watch for brown pelicans, royal terns, and Atlantic bottlenose dolphins during the summer, or scan for red-throated loons, northern gannets, and two species of cormorants—double-crested and great—in the winter. The open water of Broad Bay, fringed with a broad salt marsh, forms the park's southern boundary. The bulk of the park is inland, amidst an area of old dunes that were once along the bay's shore. Vegetation long ago colonized these old dunes, and today a dry forest of pines, oaks, and other hardwoods grows on the sandy hilltops. Drop a few feet in elevation, however, and a completely different habitat occurs. Many of the low swales between dunes lie below the water table, and freshwater swamp forest, replete with baldcypress trees and draperies of Spanish moss, grows adjacent to the dry forests. Frogs and water snakes are common in the swamps, with the cottonmouth at the northern limit of its range here. Trails meander back and forth throughout this varied forest patch, providing access to all the park's plant communities. Songbirds are common in every season in this forest, especially for a few days each fall. During the peak days of autumn migration, Seashore State Park becomes a stopping point for many migratory songbirds, a phenomenon called a *migration fallout*.

I can't discuss swamp forest in the Hampton Roads area without digressing just a bit south and west into a huge and marvelous wetland called the Great Dismal Swamp. Dismal Swamp isn't in the Chesapeake watershed despite its proximity to the lower bay, but it is one of my favorite natural history destinations. For years I was intrigued by the name, yet my first couple of attempts to explore the Dismal Swamp were disappointing. I drove all around the swamp's eastern edge, only to find a broad canal blocking all access. When I finally found roads that led to the swamp, on its western side, I didn't find the idyllic setting I had imagined—one of towering baldcypress trees growing out of still

Baldcypress trees with Spanish moss

black water and draperies of Spanish moss waving in the slightest of breezes. Instead, I found old logging roads leading through a dense growth of small- to medium-sized trees, with debris-filled ditches along the roadsides. But staring at me from the map was mysterious Lake Drummond, almost perfectly round, over 3,000 acres (1,200 hectares) in size, and far from any road. I remember walking down the trail paralleling the Washington Ditch one year in March, trudging through low, scraggly woods I thought uninteresting and tedious. I turned around after about a mile in the midst of a frigid rainstorm, wondering why I'd bothered.

I went home and did some research. Efforts to exploit the Great Dismal Swamp dated back to the eighteenth century, when a group with George Washington among its leaders attempted to drain the swamp by digging a series of ditches. The land never dried enough to be put into cultivation, but as the years went by, the forests of the swamp were gradually and systematically logged. The entire swamp had been cut by 1960, but a second-growth forest was growing back everywhere. Baldcypress and Atlantic white cedar trees were most valuable to the loggers and were most thoroughly removed. Both trees had been abundant in the swamp; now they are spotty in their distribution, though their numbers are slowly increasing again. More common trees in today's Great Dismal Swamp include red maple, river birch, tuliptree, swamp magnolia, red bay, sycamore, swamp chestnut oak, laurel oak, water oak, swamp black gum, tupelo gum, water ash, and pumpkin ash.

Although the swamp of the 1960s was no longer wilderness, it was clearly still biologically important. Several plants and animals reach their northern limit of distribution in or near Great Dismal Swamp. These include southern magnolia, tupelo gum, crossvine, green anole, yellow-bellied turtle, canebrake rattlesnake, and marsh rabbit. An entire plant community, called evergreen shrub bog or pocosin, is at its northern limit here. Animals sensitive to human activity and environmental degradation, such as black bear and bobcat, still survived in the swamp. Conservation organizations, spearheaded by The Nature Conservancy, raised funds and influenced legislation that led to the creation of the Great Dismal Swamp National Wildlife Refuge in the early 1970s. Today the refuge covers over 100,000 acres (40,000 hectares) of varied forests, swamps, and all of Lake Drummond. Even though the primeval forest is gone, the second-growth forest is slowly maturing and supports a diverse and dynamic swamp ecosystem. The old logging roads that crisscrossed the swamp are now closed to traffic, some maintained for refuge use and for recreation, others being reclaimed by the aggressive regrowth of swamp forest shrubs. The roads that are kept open are ideal for exploration, especially, I discovered, by bicycle.

There are two primary access points to the Great Dismal Swamp, both on the western side of the refuge and reached by lightly traveled backroads that lead southeast from Suffolk, Virginia. Closest to Suffolk is the Jericho Ditch area, well known to birders as an area where the regionally rare and always elusive Swainson's warbler can be found. I have found butterfly watching delightful here, with the large black and yellow palamedes swallowtail often common. The showy great purple hairstreak can often be found here, too. Watch the old canals in this area for reptiles and amphibians; spotted turtles, with lovely yellow polka dots on their upper shell, are common throughout the Great Dismal Swamp.

A little farther south is the Washington Ditch entrance to the refuge; a boardwalk trail here leads deep into the swamp forest where crossvines and yellow jessamine vines climb high into the canopy. Watch the ground for the tiny orchid called southern twayblade. Listen carefully to the warbler songs in April, as orange-crowned warbler, scarce in the mid-Atlantic, can usually be found here. These warblers migrate north quite early, however, and are usually gone before the big movement of migrant songbirds begins during the last week of April. During any day in early May more than twenty species of warblers can often be found along this boardwalk.

The main trail at Washington Ditch follows an old logging road for about 4.5 miles to Lake Drummond. It's a good dirt road that's closed to motor vehicles, making it an ideal bicycle trip. It's a rewarding journey, too, because the lake is surrounded by wild swamp forest, with some of the refuge's most mature baldcypress trees following the shallow bottom out into the lake and obscuring the shoreline in places. The water of Lake Drummond is a dark reddish-brown, stained by tannic acid released by the decomposition of leaves and other organic material. The acidity of the lake inhibits the growth of many aquatic plants; thus, there aren't many animals found here. There are some fish, though, and occasionally you'll see people fishing from a boat on the lake. Birds are never abundant on Lake Drummond, although gulls, terns, waterfowl, and osprey are sometimes present in small numbers. The trail that follows Washington Ditch to Lake Drummond can be almost symphonic in spring, thanks to the songs of hooded warbler, prothonotary warbler, yellow-billed cuckoo, eastern wood-pewee, Acadian flycatcher, red-eyed vireo, wood thrush, and ovenbird, all of which seem unusually common here.

Most visitors to Great Dismal Swamp visit either the Jericho Ditch or the Washington Ditch sections of the refuge. There are several other access points, however, and many miles of old road to explore. The Railroad Ditch, near the refuge headquarters, leads past a beautiful open wetland where beavers are active; the ditch reaches Lake Drummond in about 6 miles. This is the best-maintained road on the refuge, and there's a boat ramp at Lake Drummond at the road's end. Permits need to be obtained from the refuge headquarters to drive down this road. Travel farther south to the Corapeake Ditch, at the Virginia–North Carolina line, to see sections of the swamp where few other people travel. The best remaining examples of pocosin habitat are found along the maze of trails that connect to the Corapeake Ditch. Sphagnum moss covers the waterlogged peaty soil of a pocosin. The dominant woody plants are broad-leaved evergreen shrubs; the most common plants growing from the sphagnum are gallberry, fetterbush, leucothoe, sheep laurel, swamp magnolia, and red bay. A few pond pines are scattered here and there. Hunting is permitted in the refuge, so casual recreational travel isn't recommended during the autumn hunting seasons.

Palamedes swallowtail

Our tour of the Chesapeake country isn't finished yet, for a few outstanding natural areas on the bay's lower Eastern Shore warrant mentioning. The boundary between Maryland and Virginia on the bay's Eastern Shore lies at the mouth of the Pocomoke River. I like to call the Pocomoke our northernmost southern river because the river is lined with baldcypress swamp forest for much of its length. Many rivers in the southeastern United States feature baldcypress swamps, but this habitat reaches its northern limit at the Pocomoke. Swamps along the river can be explored at Pocomoke State Park, accessed from the west at Milburn Landing and from the east at Shad Landing. The latter area has an especially nice nature trail through the swamp, where barred owls hoot on cloudy days as well as nights. Prothonotary warblers are extremely common nesting birds here.

Farther north, at the Delaware–Maryland line, the headwaters of the Pocomoke are protected by a private conservation organization called Delaware Wildlands. A couple of old sand roads wind through this big, intact woodland, which is a mixture of baldcypress swamp and upland forest. This is a great spot for watching songbirds in spring, with Kentucky warbler, hooded warbler, worm-eating warbler, and many other forest interior birds common. The brushy edges of this forest are good birding spots too; watch this habitat for orchard oriole, yellow-breasted chat, and prairie warbler. Listen for eastern screech-owl, which is abundant here and at forest edges throughout the Eastern Shore.

Another great spot in the Pocomoke watershed lies west of the town of Snow Hill, where curiously, there are no hills and snow occurs infrequently. Nassawango Creek and its streamside swamp have been largely protected as a preserve of The Nature Conservancy, and no other section of the Pocomoke region is as pristine. The swamp forest and its rich array of plants and animals can be observed along an excellent boardwalk trail that begins near Furnacetown, where the history of an old iron furnace is on display. The land surrounding the Nassawango Swamp is sandy and covered with a dry forest, home during spring to yellow-billed cuckoo, summer tanager, and several of the uncommon little brown butterflies known as elfins. Canoe trips down the Nassawango can be delightful; check with the Maryland Nature Conservancy before boating here because there are access restrictions. The Pocomoke itself also has many ideal canoeing stretches.

Just west of the Pocomoke's mouth lies the town of Crisfield, long an economic center for the Chesapeake seafood industry. Crisfield looks a little less prosperous with every passing year, as oysters, fish, and now crabs become less available from the damaged and overfished Chesapeake ecosystem, yet the old downtown still has its charms. Ferries leave from Crisfield to visit Smith Island and Tangier Island, part of a series of small, rather isolated islands whose economies rely on the seafood industry. The marshes surrounding Smith Island and the other islands in the lower bay are lovely spots that still fill with wildlife during the warmer months. With a boat you could happily explore this section of the bay for years.

A bit north of Crisfield lie Deal Island and the towns of Wenona and Chance. The Deal Island Wildlife Management Area (which isn't on the island itself, but on a nearby peninsula) protects a vast brackish marsh and several open-water impoundments that are managed for wildlife, primarily waterfowl. Muskrats are incredibly common here, and the birding is excellent, especially during the cooler months when large waterfowl flocks are present. In many winters the regionally rare Eurasian wigeon is found at Deal Island.

The Virginia Eastern Shore is composed of many quiet rivers and marshes, surrounded by a mix of quiet forest and pastoral farms in the uplands. The natural highlights are near its southern tip in the region known as Cape Charles, where the bay reaches the Atlantic Ocean. The outstanding sites for nature study around Cape Charles are Kiptopeke State Park, the Eastern Shore of Virginia National Wildlife Refuge, and Fishermen's Island National Wildlife Refuge (see the Atlantic Coast chapter for more information about Cape Charles).

The Chesapeake Bay, including its open waters, vast marshes, and the surrounding upland habitats, is a place of great scenic beauty and tremendous ecological diversity and productivity. Sadly, this amazing natural treasure is subject to myriad abuses. As you explore some of the natural areas around the bay, I hope you'll stop and ponder the threats to this marvelous ecosystem. Consider ways you can make lifestyle changes that lessen negative effects on the bay. Many millions of people live within the Chesapeake watershed; if all of us make little contributions to improving the bay's health we will continue to benefit from its beauty and abundance.

10

Mid-Atlantic Coast

I didn't spend much time at the coast as a child. Even though we only lived about three hours away from the Atlantic Ocean, our family spent weekends and vacations visiting relatives in Baltimore, West Virginia, or the Midwest. Our few trips to the coast were almost more frightening than exciting, for I found an alien environment on the beach and a powerful, threatening force in the ocean's crashing waves. Thanks to the reassurances of my mother and father, I enjoyed the wonders of the sand, though I refused to go more than knee-deep into the ocean. Shell collecting was a marvelous treasure hunt. It was fun to watch small crabs dig in the sand and to watch flocks of little birds that my father called sandpipers running back and forth at the surf's edge.

The beach took on a different meaning by the time I reached high school.

Horseshoe crabs and shorebirds

Maryland's famous coastal resort of Ocean City was the great summer hangout for rambunctious teenagers. I heard all about trips to "O. C." and wild parties by the ocean. In typically confused teenage manner I wanted to be part of the scene, although I feared I wouldn't fit in with the party crowd. Even though I was shy and bookish, I still wanted to find fun and romance, yet I was afraid to try. Finally, when I was seventeen, I allowed some friends to take me to the beach for a party weekend. The traffic was awful. The beach was crowded and intimidating. I felt completely uncomfortable and out of place at the parties. My few attempts to find romance were painfully rebuffed. I got a terrible sunburn. I went home thoroughly disgusted with the entire concept of the summer beach resort. My friends dragged me back a few more times, which just reinforced my bad impressions. Even today I'm reluctant to go to the beach between Memorial Day and Labor Day.

A winter visit to the wild sands of Assateague Island during my college years changed my attitude. Two friends convinced me to join them for a long January hike on the Virginia side of Assateague, and the wonder of my almost-forgotten childhood trips to the beach came flooding back with the ocean waves. We had the beach to ourselves after the first five minutes of walking. Shells, including many large whelks, were scattered all along the beach. The receding tide left intricate patterns in the sand. We followed the tracks of racoons, gulls, herons, and crabs, but saw no tire tracks or human footprints other than our own. It felt as much like wilderness as any remote mountain I'd visited—and it felt a million miles away from Ocean City. I still enjoy the beach more in winter than in any other season.

Even though many mid-Atlantic beaches are overdeveloped and environmentally degraded, coastal plants and animals still thrive in many places, providing opportunities for nature study. Although the beach, dune, and marsh habitats of the coast may appear monotonous at first glance, animal and plant diversity is high. A number of beautiful wildflowers are conspicuous in all but the coldest months. A great variety of butterflies visit these flowers, including the monarchs that migrate along the coast in huge numbers during September and early October. Birding is great along the coast all year, and in every season there are rich wonders of natural history to observe. Even the landscape is dynamic, as coastal beaches and dunes are among the most rapidly moving landforms on Earth. Each season offers different rewards to coastal explorers.

I love the coast in autumn. Fall migrants (songbirds, raptors, butterflies,

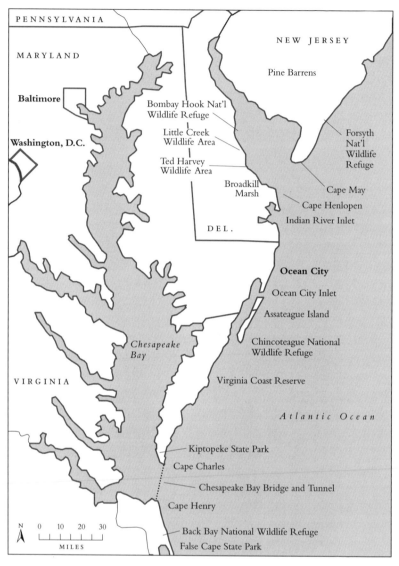

PENNSYLVANIA

MARYLAND

NEW JERSEY

Pine Barrens

Baltimore

Bombay Hook Nat'l
Wildlife Refuge

Little Creek
Wildlife Area

Washington, D.C.

Forsyth
Nat'l
Wildlife
Refuge

Ted Harvey
Wildlife Area

Broadkill
Marsh

Cape May

Cape Henlopen

Indian River Inlet

DEL.

Ocean City

Ocean City Inlet

Assateague Island

Chincoteague National
Wildlife Refuge

*Chesapeake
Bay*

VIRGINIA

Virginia Coast Reserve

Atlantic Ocean

Kiptopeke State Park

Cape Charles

Chesapeake Bay Bridge and Tunnel

Cape Henry

N
0   10   20   30

MILES

Back Bay National Wildlife Refuge

False Cape State Park

Map 6.   Mid-Atlantic Coast

dragonflies, and who knows what else) move southward along the shore, concentrating at funnel points such as the well-known and often-visited Cape May, New Jersey. Less well known is that the southern tip of the Delmarva Peninsula, Cape Charles, is an equally outstanding spot for watching dragonflies, and who knows what else) move southward along the shore, concentrating at funnel points such as the well-known and often-visited Cape May, New Jersey. Less well known is that the southern tip of the Delmarva Peninsula, Cape Charles, is an equally outstanding spot for watching the fall migration. Coastal marshes are never more beautiful than in autumn when cordgrasses turn from green to tawny brown, providing a background for the colorful impressionistic splotches painted by seaside goldenrod, sea lavender, and other late season wildflowers. Glasswort grows on the saltiest soil of the marsh, its fleshy leaves and stems turning from green to red as autumn progresses. By late fall the waterfowl show is in full swing as many ducks, geese, and swans arrive from the north to winter in our region.

There's something I cherish about the austere beauty of a sandy beach in midwinter, when icy winds blow sand into my face and shove big waves onto the beach with seemingly raucous fury. It's remarkable that many birds are found on the ocean in midwinter; if I've managed to dress warmly enough, I'll plop down onto the sand, set up my spotting scope, and scan back and forth across the water. I can almost always find red-breasted merganser, bufflehead, common loon, horned grebe, and great black-backed gull. Red-throated loon, common goldeneye, oldsquaw, the three species of scoter, and many others are also possible to see. My favorite winter finds are northern gannets, big graceful seabirds whose nesting colonies in Atlantic Canada I long to see. When winds blow from the east I sometimes see thousands of gannets cruising the nearshore waters, gliding gracefully and then crashing into the sea with spectacular free-fall dives.

Onto this palette add the wind-up toy antics of sanderlings, which seem to chase the waves up and down the beach, and the arresting beauty of a snow bunting flock moving through the dunes. Completing the picture are the austere beige of the sand, the warm tan of the beach grass, the dark green and brown of wax myrtle groves, the deep blue of the ocean, and the ominous gray of a winter sky.

Spring is an enchanting time in every section of the mid-Atlantic. Many naturalists eagerly await the first wildflower blooms of the season, the northbound movement of migratory birds, and the return to activity of winter-dormant reptiles, amphibians, and butterflies. Because of the moderating influence of ocean waters on coastal climates such spring signals arrive at the coast before they reach inland sites. By early March pine warblers inaugurate the songbird nesting season as they sing from territorial perches in the maritime forests that lie just inland from the beach. Waterfowl migration begins about the same time, followed by the migrations of raptors and shorebirds. Early spring wildflowers pop up here and there near the coast; if you're watchful you may find pink ladyslippers beneath coastal pines in April. The first flush of new green growth in the salt marshes coincides with the flight period of early season butterflies. One of our region's most impressive natural phenomena occurs in May, with the synchronous mating of millions of horseshoe crabs. As sheltered beaches along Delaware Bay and the backsides of barrier islands become crowded with egg-laying horseshoe crabs, great flocks of shorebirds descend to feed on the nutrient-rich eggs. These intercontinental migrant shorebirds are at their most beautiful in May, when they are freshly molted into their breeding plumage. You can't go wrong on a springtime visit to the coast.

Even though summer at the coast can mean traffic jams on the roads, crowds on the beach, higher prices for lodging and meals, and troublesome biting insects, it is also a season of great natural activity. Even on the most crowded beaches nature keeps chugging along in spite of the crowds and the disturbance. Summer is the season of greatest wildflower abundance and variety in many coastal habitats. Butterflies, including some that can rarely be found in other parts of the region, occur in great numbers and diversity along the coast on the warmest sunny days. Wetlands are awash with the fecund riot of marine and estuarine life, as those who try to catch finfish or blue crabs are well aware.

Fall migration along the coast actually begins in early July, when shorebirds and terns start to return from northern nesting areas. Most shorebirds migrate in two distinct waves: The adults head south shortly after their eggs have hatched, leaving the young with sole access to the food resources of the nesting region. In late July and early August virtually all the migrant shorebirds in our region are adult birds, many still in their showy breeding plumage. By early September, though, juvenile birds predominate, offering yet another set of identification challenges. Mixing with the birds heading south at midsummer are ones who breed to

Dragonflies

our south and which sometimes wander north into our region after their nesting season is over. Midsummer is the time we're most likely to find such southern strays as limpkin, white ibis, or reddish egret.

Let's travel now to important natural areas along the Atlantic coastline from southern New Jersey down to the Virginia–North Carolina border. We'll take a side trip up Delaware Bay to discuss some of the important natural areas found along this estuary.

NEW JERSEY

Cape May is a lively resort community at New Jersey's southern tip; it is a town well known for its Victorian-era architecture and ambiance. Naturalists look at the map of New Jersey and see how a broad area of land gradually tapers southward to the narrow Cape May peninsula. Many birds prefer to migrate over land rather than over water, so as they head south over New Jersey they are funneled together into increasingly smaller land areas the farther south they go. This funneling culminates at Cape May, where the concentration of songbirds, shorebirds, and raptors during the fall migration is legendary. Catch the weather and season just right, when favorable conditions allow huge numbers of birds to be on the move, and birding at Cape May is extraordinary. What are the ideal conditions? Nobody knows for sure, but the passing of a cold front usually increases the numbers of birds seen at Cape May, whereas low clouds and rain bring migration to a halt.

Autumn at Cape May is known worldwide as a great birding spectacle, and during autumn weekends there are incredible concentrations of birders to go along with the birds. As a result, any rare bird that shows up at Cape May is likely to be found by some birder, and word about it spreads quickly. If your goal is to see a lot of birds, both total numbers and variety, and to have a chance to see rarities, Cape May in the fall is hard to beat. You'll find that the counters at the official hawk watch are also tallying the butterflies and dragonflies that zip by. Add fall wildflowers to the mix, along with the town's pleasant setting and its abundance of fine hotels and restaurants, and it's easy to understand Cape May's popularity.

A couple of other destinations, on or near the New Jersey coast, are within reach of Washington, D.C. and Baltimore for weekend exploration. The Edwin B. Forsythe National Wildlife Refuge (most people still call it Brigantine, its previous name) preserves extensive salt marsh, tidal creek, and beach habitats north of Cape May. Huge flocks of snow geese frequent Brigantine's marshes from October to March, when many other species of waterfowl may also be observed. Highlights of other seasons include shorebirds, fiddler crabs, butterflies such as the salt marsh skipper and the buckeye, and other insects, most notably (alas) ferocious salt marsh mosquitoes.

Just inland from the southern coast of New Jersey is one of the most distinctive natural areas I've ever explored, the New Jersey Pine Barrens. The land here is essentially a huge sandpile underlain by a massive freshwater aquifer. The water table breaks the surface at the lowest points within the Pine Barrens; the swamps here are dominated by Atlantic white cedar, and there are extensive bogs where cranberry, pitcher plant, sundews, bog orchids, and curly-grass fern grow. The rivers of the Pine Barrens are narrow but fairly deep, generally without rapids yet with steady flows. The acidic water is tinted a dark reddish-brown color from tannins produced by the decomposition of plant materials. Canoeing the beautiful rivers of the Pine Barrens, such as the Batsto, Oswego, and Mullica, is an enjoyable way to see the area's wetland habitats, and the many canoe rental outfitters in the region attest to the popularity of this activity. Spring and fall are less crowded seasons than summer for a trip on one of the Pine Barrens' rivers.

Most of the Pine Barrens region is above the water table; thus, the sandy soil retains little moisture between rains and never holds many nutrients. The result is the rather barren habitat that gives the region its name. In some places pitch pine forests grow, and in other places scrubby deciduous trees dominate, with blackjack oak and post oak most common. None of the pine barren forests ever have many tall trees, in part because of the poor soil and in part because of the region's susceptibility to fires. Where the conditions are harshest, a dwarf forest develops; it's an eerie feeling to stand on a little knoll and look out across vast forested areas where the trees average only about 4 feet in height.

The Pine Barrens support a distinctive wildlife community. The wetlands are home to many turtles, snakes, and frogs, including the endangered Pine Barrens treefrog. Upland forests fill with prairie warblers, eastern towhees, and whip-poor-wills in spring and summer. Explore the uplands on foot trails or from the maze of sand roads that wind back to places that feel remote from modern civilization. When you're way back on one of these roads, feeling a little lost and worrying about whether your car will make it through the sand without getting stuck, it's easy to understand why the Pine

Barrens have generated more than a few folktales. Lore of the Pine Barrens includes tales of ferocious packs of feral dogs and a horrible, mythical creature called the Jersey Devil. At least I think it's mythical. . . .

DELAWARE BAY

Most birds that gather at Cape May either double back and hug the shores of Delaware Bay, or they make the flight over the mouth of the bay, crossing over to Cape Henlopen, Delaware. Cape Henlopen is a state park whose best-known feature is its long, active sand spit that is changing about as rapidly as any landform in the mid-Atlantic. Various longshore currents move tremendous volumes of sand around Henlopen's point. Several jetties and breakwaters that were built in the historic era have dramatically altered the sand flow. The result is a huge sandpile that is extending out into the sea on one side, rapidly eroding on another, and building up into impressively large, fast-moving dunes just inland. An old military tower is now open (on days when the weather is dry and calm) as a scenic viewpoint, from which all of this dynamic landscape may be observed. The entire cape was an active military base during World War II; pamphlets are available at the park's visitor center that describe Henlopen's interesting history.

Cape Henlopen has many migrating hawks and shorebirds, as well as a wide assortment of nesting birds in summer, but to me Henlopen is a winter place. The sparsely vegetated dunes on the cape are home to large wintering flocks of snow buntings, along with smaller numbers of Ipswich sparrows and, in many winters, a snowy owl or two. Red-throated loon, great cormorant, and peregrine falcon are often seen. I still marvel at my memories and photographs from a frigid winter's day visit to the tip of the cape, where big chunks of ice had piled up along the beach, stacked up 10-feet high in places. It looked like the Arctic Ocean! Cape Henlopen State Park can be crowded in summer, but there's plenty of elbow room in winter. Still, during warmer weather life is much more vibrant along the coast, and Henlopen is worth visiting at any season. Fishing is excellent from a long pier near the park entrance. You can study songbirds, butterflies, and wildflowers near the campground. The quiet stretch of beach on the bayside of the point is a rewarding area for beachcombing. Watch for closures of certain areas for the nesting of the threatened piping plover and least tern. The park is just south of the Cape May Ferry dock near Lewes, Delaware.

Birds that don't cross the broad expanse of Delaware Bay at its mouth are likely to follow its shoreline, and fortunately there are many protected natural areas on both the New Jersey and the Delaware sides. The Delaware sites, being much closer to my home, are the ones I usually visit. Delaware Routes 9 and 1, along with U.S. Route 113, parallel the bay's shore for much of its length and provide access to many wetland areas. Some of the better known areas, from south to north, include Broadkill Marsh, Prime Hook, Slaughter Beach, Ted Harvey, Kitts Hummock, Little Creek, and Bombay Hook. Many of these places are horseshoe crab hot spots, where these primitive arthropods occur in staggering abundance during the latter half of May, laying eggs that in turn become food for hungry shorebirds. Many shorebirds migrate from the South American coast up into the high Arctic; without a small handful of spots where food is abundant, such as Delaware Bay at horseshoe crab egg-laying time, this long-distance migration would be impossible.

The Delaware Bay shore has little sandy beachfront and no resort communities, so even in midsummer this area is pleasantly uncrowded. The heat, humidity, and biting insects are here, but in late July and August so are many shorebirds, terns, herons, egrets, butterflies, dragonflies, and wildflowers of the marsh and meadow. Bring insect repellent, sunblock lotion, and a big jug of ice water along with your binoculars to enjoy a summer visit to the Delaware Bay shore. In winter dress warmly, bring hot drinks, and be prepared for colorful sunsets and amazing wildlife scenes.

Broadkill Marsh is just a few miles north of Cape Henlopen along the western shore of Delaware Bay. I always end a winter day's excursion to Henlopen with a visit to Broadkill, a section of the Prime Hook National Wildlife Refuge. This vast marsh is beautiful at any season, but from November through February the crisp, cold air of winter almost always makes the sunset richly colorful, an excellent backdrop for the twilight flights of snow geese. Many thousands of these lovely black-tipped white geese fly from the surrounding countryside into the marsh during the sunset hours. It's an incredible scene to watch as six or eight groups of hundreds of geese each drift against the orange western sky, dropping down into the marsh one group at a time. Some always seem to fly directly overhead, their incessant honking adding music to the scene. Numerous Canada geese are present, and small groups of ducks also zip by, looking like whir-winged bullets compared with the more leisurely flying geese. Watch the edges of the marsh for red fox, gray fox, or white-tailed deer. Near the eastern end of Broadkill is a big section of open water where a good variety of ducks may usually be observed. Muskrats swim silently across the water while the marsh grasses in the background gently sway in the wind. Where northern harriers hunted over the marsh during the daylight hours, watch for their nocturnal counterparts, short-eared owls, late in the afternoon and through the twilight. Listen for the calling of great horned owls as sunset's last glimmer fades away.

Another popular field trip to the Delaware Bay shore features visits to the Little Creek and Ted Harvey Wildlife Areas. There are three primary access points to the Little Creek Wildlife Area, which is a series of manipulated wetlands managed by the state. The northern access is Port Mahon Road at the tiny hamlet of Little Creek, just east of Dover. This road reaches the Delaware Bay shore by the mouth of a little tidal creek. There's no formal parking area here, but on weekends during May you'll surely find cars parked at the road's edge. Horseshoe crabs gather along the shores of this creek in astonishing numbers; at times the creek edges are covered with horseshoe crabs, mating and laying eggs, with a glimpse of the mud flats barely visible. I remember a time, not long ago, when these animals were given no legal protection, and it was common to find people filling up the backs of big trucks with living, writhing horseshoe crabs. Although some of these crabs were used for biological or medical research, huge numbers were just ground up for fertilizer. Thankfully, their mass collection is no longer legal. In May there are always lots of shorebirds at Port Mahon, taking advantage of this abundant food resource. At the mouth of the creek you can see many of these birds at close range, which is a nice bonus for experienced birders and an essential requirement for beginners trying to sort through the many look-alike species. Stay alert for other birds at this spot, too. Clapper rails and seaside sparrows are common here, and peregrine falcons, ospreys, and other raptors may often be seen overhead. At low tide the pungent scent of the salt marsh fills the air. Long ago I found this smell unpleasant, but now the fragrance is one I enjoy because it connects to so many great memories.

Other access points to Little Creek lie south of Port Mahon and offer views of shallow freshwater impoundments and marshes. These areas fill up with shorebirds, gulls, terns, herons, and egrets during the warmer months and with waterfowl and gulls in the winter. A little farther south, the Ted Harvey Wildlife Area, another state facility, protects similar habitat. Continue past the entrance road to Ted Harvey and visit the bay's shore at Kitts Hum-

Seaside sparrow

mock, another spot where horseshoe crabs gather in unbelievable densities during late May.

Little Creek and Ted Harvey are especially popular among experienced birders because each area has a history of harboring rare birds, especially during the early part of fall migration, late July and August. In recent years these areas have been visited by whiskered tern (the first time this species was recorded in North America), white-winged tern, reddish egret, white ibis, curlew sandpiper, red-necked stint, ruff, and other birds that are rarely seen in the mid-Atlantic states. When one or more of these rarities are present this becomes a great spot for birder watching, attracting some of the region's best-known and most experienced birders. If you're new to birding, it can be rewarding to visit a spot that is so popular because experienced birders are generally delighted to share their finds and knowledge with beginners. Just be courteous and don't interrupt a formal field trip with more than a few questions.

Bombay Hook is a national wildlife refuge along the Delaware Bay shore northeast of Dover. A popular wildlife drive provides access to freshwater impoundments and the seemingly endless salt marshes that line the bay. Bombay Hook features opportunities to see many of the birds listed for Broadkill, Little Creek, and Ted Harvey, though there is no access to the bay's shore itself from the wildlife drive, so you can't see the great horseshoe crab concentrations here. Unlike the other areas, though, Bombay Hook offers access to significant upland habitat, including deciduous forest, brushy meadows, and agricultural fields. This makes Bombay Hook an excellent place to search for butterflies, mammals, and upland birds such as ring-necked pheasant and eastern meadowlark.

THE ATLANTIC COAST: DELAWARE THROUGH VIRGINIA

South of Cape Henlopen along the Delaware shore is the resort stretch of sandy oceanfront, with a little open dune and coast remaining between Dewey Beach and Bethany. These communities are busy places in summer, and every year brings more building and less natural habitat to this region. Still, there are a few natural surprises here, with the Indian River Inlet my personal favorite. This is a great spot in winter, when purple sandpipers dot the rock jetties and sea ducks, including the occasional eider or harlequin duck, bob alongside loons and cormorants in the turbulent sea. A Ross' gull was found here in November 1996. Check out the engineering device that pumps sand through a tube on the highway bridge, hopefully minimizing the need to dredge the boating channel through the inlet in future years.

Development has destroyed a lot of habitat along the southern Delaware coast, but cross into Maryland along the coast road and wildlife habitat virtually disappears. This is Ocean City, where summer brings an incredible concentration of wild life, but not much wildlife. There's not much room for birds on the Ocean City beach, though gulls manage to thrive even with the crowds. There are a few highlights, though. The city covers a narrow strip of land between the Atlantic Ocean and Assawoman Bay. A narrow, sandy island in the bay known as the Fourth Street Flats can be observed from the western end of Third, Fourth, or Fifth Streets in Ocean City. Birds frequent this island though they are always fairly far away, but a spotting scope will often reward the viewer with sightings of black skimmer, brown pelican, great blue heron, and many different shorebirds, gulls, and terns in the summer. Winter birds of this island and the surrounding waters may include brant, red-breasted merganser, and American oystercatcher. Assawoman Bay is an excellent fishing spot, and people often line the Route 50 bridge during summer dropping lines into the channel just south of the flats. I remember one time when a big school of bluefish passed under the bridge. Friends and

I couldn't drop a hook into the water without catching another fish almost instantly.

The Ocean City Inlet lies at the southern tip of the town, and here the nutrient-rich bay waters mix with those of the open ocean. The result is a fertile marine environment that is good for birds even though it occurs at the city's edge. It's a difficult area to access in summer, with all the crowds coming to visit the boardwalk, amusement park, and the countless shops of this area, but once the weather turns cold this is an excellent birding stop. At times the inlet is a flurry of bird activity, as hundreds of common terns or Bonaparte's gulls may sometimes be found here. There's a long list of possible finds at the inlet in winter, ranging from northern gannets cruising the ocean to a peregrine falcon that likes to sit on the nearby municipal water tower. During the warmer months watch for Atlantic bottlenose dolphins, which sometimes swim through the ocean past the inlet.

South of Ocean City the coast takes a wilder turn. Assateague Island stretches for about 40 miles, and although certain areas of the Assateague beachfront may be crowded with recreational beach activity, many other areas can only be reached with a significant hike. Even in midsummer there aren't many people in these areas. Assateague's beaches teem with marine life, with the secretive, pale ghost crabs among my favorites. Interesting shells sometimes wash up onto the beach here, with the big whelks especially popular among beachcombers. Shorebirds frequent Assateague in all seasons, and in winter the offshore waters can hold many rarities. Roads on the northern section of Assateague lead to a Maryland state park and a federal facility, the Assateague Island National Seashore.

Hognose snake

Oversand vehicles are permitted on the beach from here south to the boundary with Virginia, but vehicles are prohibited from the state park north to the Ocean City Inlet. I love to walk this wild, windswept section of the island. Jetties at Ocean City block the southern flow of sand offshore, and as a result the northern section of Assateague doesn't receive regular replenishment of sand. The island is eroding away! While it's still here, however, the northern section of Assateague is a mix of dunes and shrubs, backed by salt marsh on the western edge. It's almost 10 miles round-trip to walk to the island's northern tip and back. As a result, this section of the island is rarely visited and maintains a wild, primal feel. Well, it maintains this feel until you look north and see the roller coaster, Ferris wheel, and high-rise hotels and condominiums of Ocean City.

Farther south, where Assateague Island is wider, you find loblolly pine forests, dense wax myrtle shrub stands, and other upland habitats. These areas are excellent spots to watch for songbirds in spring, summer, and fall, and to seek the tiny, elusive northern saw-whet owl in winter. Watch for the comical hognose snake on warm days. This harmless snake will hiss, flatten its head, and wave its tail ominously when approached, trying to make you believe it's a rattler. If that fails, a hognose will often roll over and play dead, emitting a putrid musk. Hognose snakes come in two color forms, one patterned dark brown and yellow, the other uniformly dark gray. Summer and fall bring many wildflowers to the transition zone where these upland habitats merge into the salt marsh, along with a great assortment of dragonflies, damselflies, and other interesting insects.

The southern portion of Assateague Island is in Virginia, and here the island is several miles wide in places. Extensive maritime forests, some uniform stands of loblolly pine, and other stands with mixtures of pines and a variety of hardwoods, occur on southern Assateague Island. The entire Virginia section of Assateague is protected as the Chincoteague National Wildlife Refuge. Although significant recreational activity is permitted on the refuge, access is limited in many areas; as a result, this is an outstanding wildlife area. The beach may be accessed near the Tom's Cove Visitor Center, a facility operated by the National Park Service. This is a popular area for swimming, surfing, fishing, kite flying, and sunbathing; thus, it is not a good beach for wildlife. Still, many birds fly over the ocean here, especially during migration. From about a mile south of the visitor center to the island's southern end, an area known simply as The Hook, all access is restricted from March through August for the protection of endangered beach-nesting birds. During the cooler months access to this area is permitted by foot, boat, or oversand vehicle.

From the Tom's Cove Visitor Center north to the Maryland line, about a dozen miles, only foot access is permitted. After the first mile you'll find few other people along this wild and remote stretch of beach. Several times I've made the long walk up this beach all the way to Maryland, where camping is permitted. I've usually gone during late fall or winter, when I've had the whole section to myself, and during the full moon, when the tidal range is extreme and the nights are awash with enchantment. My legs and feet always grow sore because the sand makes walking arduous, but my spirit is always refreshed from the beauty of the coastal scene, the richness of the wildlife, and the peaceful solitude of an area seemingly untouched by humans. As I walked this beach one Thanksgiving day I saw a raccoon out in the shallows digging for crabs or other sand dwellers, oblivious to the small waves breaking over its back. My photographs of "surf raccoon" always draw a few chuckles.

Behind the beach, the Chincoteague National Wildlife Refuge offers scenic drives and trails that pass through Assateague's varied forest and shrub communities and past freshwater pools and marshes, open fields, and salt marshes. Birding is great in all seasons, though in dry years the freshwater areas can disappear by late summer and the numbers of birds drop. Not only are waterbirds abundant, but songbirds also thrive in the upland habitats and many raptors migrate overhead, especially during early autumn. Because of its great habitat diversity and its protected status, the Chincoteague refuge merits a long visit by any coast-loving naturalist. At Chincoteague I saw my first northern gannet, my first black-headed gull, and my first salt marsh skipper, a dainty little butterfly that darts from flower to flower in marshy areas during summer and fall. On the beaches I've discovered many fascinating marine creatures washed ashore, ranging from the elegant forms of big whelk shells to the peculiar pink blubbery masses known as sea pork, which are groups of colonial protochordate marine animals.

South of Assateague Island a long stretch of barrier islands fronts the Atlantic, most of which are not readily accessible to the public. Many are protected by The Nature Conservancy as its Virginia Coast Reserve, which ensures that they will be protected and managed for optimal wildlife use. Selfishly, I'd love to see these places, but intellectually I'm glad that some places are left undisturbed as relatively intact ecosystems. The Nature Conservancy does sponsor a few trips out to sections of the reserve each year for members and for specialized groups, including some school classes.

Down at the southern tip of the Delmarva Peninsula—the long expanse of land lying east of the Chesapeake Bay in Maryland, Delaware, and Virginia—is a region known as Cape Charles. Don't confuse the town of Cape Charles, almost 10 miles to the north, with Cape Charles itself, which is the peninsula's tip. Here the Cape May phenomenon is repeated, as a broad expanse of land funnels migrating songbirds, shorebirds, raptors, and butterflies into a narrow point of land. Catch the weather and the season just right and Cape Charles is an amazing place, with peregrine falcons darting overhead, warblers filling the woods, and thousands of monarch butterflies flitting around every open field. For years Cape Charles was a well-kept secret, but when massive development projects were proposed for the area in the 1980s, the conservation community mobilized a successful battle, which resulted in a new Virginia state park called Kiptopeke. At the same time visitor access and services were greatly improved at the adjacent Eastern Shore of Virginia National Wildlife Refuge. Cape Charles is no longer secret, but it's never overly crowded, even over the early October weekend when the Eastern Shore Birding Festival is held here. Walk trails through the many habitats of Cape Charles and you'll surely see a great variety of plants and animals. Want to see animals up close? Go to Kiptopeke State Park in the fall, when you can visit a songbird banding station in the woods and a raptor banding project in the open fields.

Cape Charles is the northern terminus of the Chesapeake Bay Bridge and Tunnel, an amazing structure that allows one to drive across the broad mouth of the Chesapeake Bay. The two tunnels are anchored by four artificial islands, three of which require special permission to visit. All can be outstanding birding spots; it's almost like driving out into the ocean. Seabirds are the highlight, especially during the winter months, but every migration season brings a few surprises to the islands. The city of Virginia Beach lies at the southern terminus of the bridge and tunnel, and three more coastal areas of note are found here. Just east of the bridge is Seashore State Park, which protects an unspoiled section of beach and dune just west of Cape Henry, the southern side of the Chesapeake Bay mouth. An old dune system at the Seashore State Park is crisscrossed with walking trails that lead through a fascinating set of habitats, including dry sand ridges and blackwater swales, where baldcypress trees dominate. See the Chesapeake Bay chapter for more about Seashore State Park.

South of Cape Henry is the oceanfront resort development of Virginia Beach, a poor area for wildlife. Farther south, though, are two excellent natural areas, the Back Bay National Wildlife Refuge and False Cape State Park. Back Bay includes beach, dune, and freshwater habitats and is a vital resting area for many migratory birds. False Cape is a remote park spanning the coastal strip from the southern edge of Back Bay to the North Carolina border. There's a beautiful, remote stretch of beach at False Cape that's often teeming with wildlife, though sadly a handful of nearby residents are still permitted to drive oversand vehicles on this beach. Inland False Cape protects a fairly mature section of maritime forest—woodlands that can be filled with songbirds during spring and fall migration. It's a great place to study coastal plants of forest, field, dune, and wetland habitats, and to watch for butterflies such as the great purple hairstreak. Access to False Cape State Park is limited: Either travel by boat or overland on foot or bicycle across the Back Bay National Wildlife Refuge. Access through Back Bay is limited to the sandy beach (which is not good for bicycling) during the winter and early spring months, in order to minimize disturbance to the birds that winter on the refuge impoundments.

Here ends our tour of the mid-Atlantic coast. If we were following migrant shorebirds south we could continue along North Carolina's Outer Banks, visit Huntington Beach and Cape Romain in South Carolina, Cumberland Island in Georgia, and any of a great number of spots on the long Florida peninsula. Next we'd head across the Caribbean, perhaps stopping at one or two islands before landing on the Venezuelan coast, then continuing on down to Brazil and Argentina. We'd retrace our steps in spring, continuing north from the mid-Atlantic to Sandy Hook, New Jersey; Montauk Point on Long Island; Cape Cod, Massachusetts; and all the way north to shores of the Arctic Ocean on Baffin and Ellesmere Islands. Shorebirds have linked coastal environments from the Arctic to the Antarctic for many centuries; perhaps they'll inspire us to link these areas in a comprehensive, hemispheric conservation plan.

11

Close to Home— Nature near Baltimore and Washington, D.C.

I'm a child of the Maryland suburbs. I was born in Baltimore and raised in the suburbs half a dozen miles north of Washington, D.C., in a development built during the mid-fifties on top of productive farmland. An old farmhouse still stands a couple blocks from my childhood home; I often tried to imagine how it looked as a farm.

My early experiences with nature were spotty at best. In one of my earliest memories I walked with my grandfather past a corner house with hickory trees in the backyard. My grandfather knew everyone in the neighborhood, and as we stopped to chat he reached down and handed me a hickory

Black-crowned
night-herons at the
National Zoo

nut still inside its four-part shell. It was a delightful little puzzle to pull the shell apart and then put the pieces back together. Each nut was a little different, a variation on the same challenge. Pieces from one shell wouldn't fit around another nut.

A little later friends introduced me to the small patch of woods behind our church, where a tiny stream flowed. I watched with awe as they lifted the stream rocks and caught crayfish in paper cups; I was afraid to try, frightened by the big pinching claws of those monsters! Eventually I gained the courage to handle the crayfish and the salamanders we'd sometimes find.

As I grew older my family sometimes went on the great excursion up to *The Park*. Just a couple miles from home was Wheaton Regional Park, whose primary attractions were a miniature train to ride, a playground featuring an old jet we could climb and play on, and at *Old MacDonald's Farm*, a small collection of farm animals to see and pet. Beyond the developed area was a patch of woods so large as to seem endless, with trails leading past creeks, over hills, and even to a lake. Friends and I were allowed to explore only so far, being gravely warned about the dangers of getting lost in this wild place.

By my teenage years I'd made trips to the C & O Canal, the Catoctin Mountains, and even along the Appalachian Trail, and I looked back disdainfully at my neighborhood parks. There's nothing there, I mused; how unfortunate I thought I had been to grow up with nothing better than that to explore. As a teenager I wondered why our family couldn't live out in the country. Why couldn't *we* vacation in the great national parks out west? As I began to care more and more about wild places where nature reigned, I grew to hate the suburbs and dream of the day I could move away.

Over time I was able to explore West Virginia, the Great Smoky Mountains, Maine, Vermont, Florida, and then the American West. Although I was overwhelmed by the grandeur of these exotic places time and time again, my attitude toward the parks back home slowly began to change. Of course there was no Mt. Katahdin in Wheaton, no Everglades in Alexandria, and my friends in Seattle would never agree to my proposed trade of Tyson's Corner for Mount Rainier, but I began to realize that every little patch of nature has value and wonder. I latched onto a handful of spots close to home where I'd walk several times a week. Woods along the Potomac River and the Northwest Branch sustained me through graduate school, though they might have delayed the completion of my degree by several years.

As our cities and suburbs grow, I watch as outlying natural areas are over-taken by the bulldozer and, even more horrifying, as many of the little fragments of nature tucked in close to town are also being developed. It's these small but accessible natural areas that often play vital roles in the lives of young people, introducing them to the wonders of nature. Thankfully natural habitats are protected with a diverse set of parks close to Washington, D.C. and Baltimore. I continue to visit some of these local parks regularly, and I'm often amazed at the natural wonders that I find intact. They may not be grandiose, but our nearby fragments of nature are worthy of protection and respect.

Any natural area close to home is worth visiting regularly, no matter how small and ordinary it may seem. Repeated visits to such places provide a window onto the progression of events through our yearly cycles. When teaching I've often asked students to observe a single tree at least once a week, noting any changes they see in the tree and its surroundings from week to week. Whether observing a single tree or a forest ecosystem, I find that nearby spots I can study regularly teach me things that exotic locales I see infrequently never could.

Not every spot close to Washington, D.C. or Baltimore is abused and or-dinary. Another benefit of travels far afield is perspective. There is biological richness and diversity within the urban and suburban parks of the mid-Atlantic that rival some of the wilder and more grandiose corners of the country. Fortunately we have protected large sections of the valleys of most of our rivers and streams, even those running close to our cities. Most of the parks and other protected natural areas in the Washington-Baltimore area are found in these valleys, though I am grateful that a few upland areas are also preserved.

The following overview of natural areas found close to the cities of Wash-ington, D.C. and Baltimore is organized by watershed. Maryland areas are discussed first, moving from the west to the east. Areas in the District of Co-lumbia are included with the Rock Creek and Anacostia sections. The chapter concludes with descriptions of sites in Virginia.

POTOMAC RIVER

Parks along the Potomac River are some of the best in the entire region and have been among my favorite spots to visit for more than two decades.

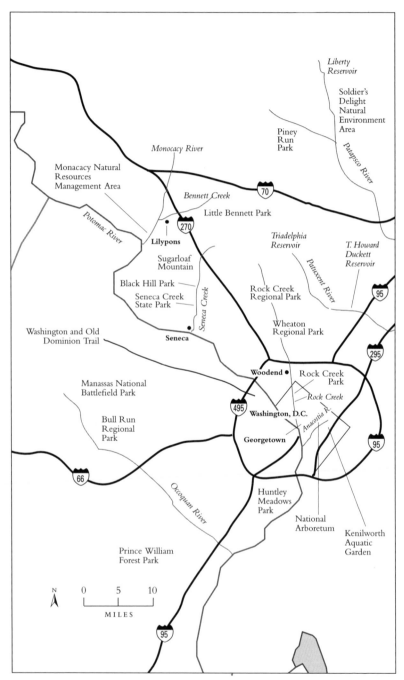

Map 7. Close to Home

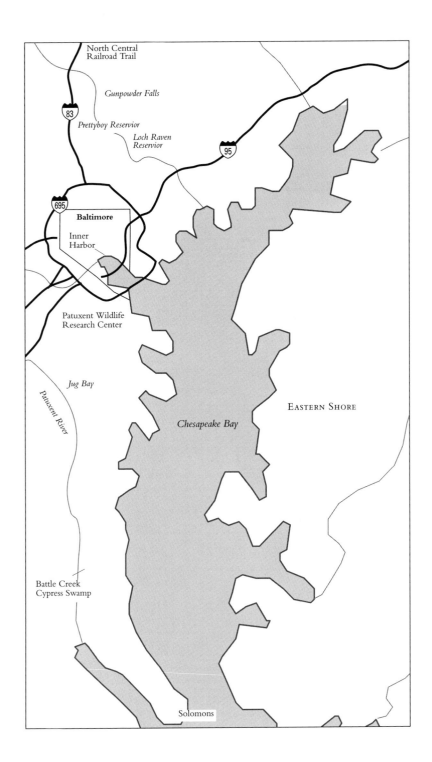

North Central
Railroad Trail

Gunpowder Falls

83

Prettyboy Reservior

*Loch Raven
Reservior*

95

695

Baltimore

Inner
Harbor

Patuxent Wildlife
Research Center

Jug Bay

Patuxent River

Battle Creek
Cypress Swamp

EASTERN SHORE

Chesapeake Bay

Solomons

Notable natural areas close to Washington, D.C. along the Potomac, on both Maryland and Virginia shores, include the C & O Canal, Hughes Hollow, Riverbend Park, Great Falls, Dyke Marsh, and Mason Neck. Each of these areas is described in the Potomac chapter.

MONOCACY RIVER

The Monocacy River drains much of Frederick County, Maryland, as the river flows south through the pleasant rolling farm country of the Piedmont, passing east of the town of Frederick en route to the Potomac. Declared a state scenic river, sections of the Monocacy are good places for canoeing when the water levels permit. There's not much parkland along the river itself, though the Monocacy National Battlefield is just south of Frederick. Although the battlefield is primarily a park preserving Civil War history, it is located at a peaceful place along the river bank. Open fields here fill with butterflies in the summer and with goldfinches in the fall. Closer to the Potomac is the Monocacy Natural Resources Management Area, where trails lead through pleasant floodplain forest and past brushy meadows where blackberries and wineberries grow. This is a popular hunting area; I don't recommend autumn visits.

A short distance upriver from the Monocacy Natural Resources Management Area, the Monocacy is joined by Bennett Creek, a significant tributary flowing from east to west. Near the headwaters of a creek that flows into Bennett Creek is an outstanding natural area, Little Bennett Park, which is named after this creek. Little Bennett Park features a low valley surrounding the creek where an active beaver colony maintains a mix of ponds, swamp, marsh, meadow, and forest. Maryland's official state insect, the Baltimore checkerspot butterfly, is rather scarce in the state, but it may be found at Little Bennett. Turtlehead, a wildflower in the snapdragon family, is the primary food plant for Baltimore checkerspot caterpillars. Turtlehead is uncommon across the state, but it grows in wet meadows next to Little Bennett Creek, and the checkerspot can be seen flying here each June. The state's official bird, the Baltimore oriole, is usually easier to find than is the checkerspot, both at Little Bennett and across the state. Orioles build their woven, pendulous nests in large trees with spreading crowns. These birds migrate to tropical America in late summer, returning in late April the next year.

My favorite stretch of Little Bennett Creek lies south of Clarksburg Road, between the little parking lot and the old, fenced-off Kingsley Schoolhouse, a historic one-room school. The brushy creek valley near the trail's beginning is home to nesting blue-winged warbler, yellow-breasted chat, willow flycatcher, and many other birds. Where the woods close around the trail the list of nesting birds expands to include worm-eating warbler, Kentucky warbler, yellow-throated vireo, and scarlet tanager. Butterfly diversity is often high along this stretch, with coral hairstreak, snout butterfly, and great spangled fritillary as easy to find here as anywhere near Washington, D.C. This stretch of the Little Bennett Valley is a great place for amphibians, too, with a high variety of frogs and salamanders found here. I once rolled over a log at the trail's edge and found a long-tailed salamander, a species said to be very rare in the mid-Atlantic outside of the mountains.

Other trails around Little Bennett lead through rolling forests and past old fields being reclaimed by forest. Red cedar trees are common in some of these old meadows, and in July these trees harbor numerous olive hairstreaks, attractive little green butterflies. Topography doesn't suggest that Little Bennett lies on any hawk-migration corridor, yet every time I visit the park in autumn it seems like I see an assortment of southbound hawks overhead. Like most natural areas listed in this chapter, Little Bennett has had to survive many threats to its integrity. First was a plan to dam the creek and create a lake solely for recreation; this plan was defeated. Then came a plan to build a golf course in the middle of the park, which would have broken the extensive forest block that is vital to the park's biological richness; the golf course was built, but at the edge of the park, not in the center. An elaborate conference center has been planned for the park, and although it hasn't been built and the plan has been shelved, it could resurface some day. Those of us who cherish local parks need to constantly speak up for their protection.

Shortly before Bennett Creek reaches the Monocacy, it swings around a great loop along the northern base of one of our region's best-known natural landmarks, Sugarloaf Mountain. Sugarloaf's summit is 1,282 feet (390 meters) above sea level, hardly one of the world's great mountains, yet it's the highest point for miles. Most of the Piedmont region has eroded to much lower elevations, the mountains formerly found here worn down to a low, rolling landscape. A dense cap of erosion-resistant quartzite, about 200 feet (60 meters) thick, kept Sugarloaf from wearing down along with its surroundings. Isolated mountains formed in this manner are called monadnocks. Roughly 3,000 acres (1,200 hetares) on and around the mountain are part of a private reserve called Stronghold, the old estate of Gordon Strong. Pleasant trails, open to the public during daylight hours, lead to the mountaintop through the upland forest. This habitat isn't as rich as those of the surrounding stream valleys, yet many plants and animals are found here. During spring migration, Sugarloaf is an excellent spot for warblers and thrushes, with magnolia warbler and black-throated green warbler particularly fond of the summit area. A pair of common ravens has been nesting at the mountain for years, the only spot east of the Blue Ridge in our area where you can enjoy their croaking calls and aerobatic flights. Parts of the Stronghold property are logged each year, but extensive forests remain, some quite mature and harboring wild orchids and other rare wildflowers.

A private road leads through the Stronghold property almost to the summit of Sugarloaf. The road and the mountain's trail system are free and are open during daylight hours year-round. The road is busy on nice weekends; go early in the morning or on a weekday if you can, especially during spring and fall. If you follow the trails away from the summit, however, you'll quickly get away from the crowds. Most of Sugarloaf's trails begin along the road to the summit, but the trail system can also be accessed from the gravel Mt. Ephraim Road, which winds through fairly mature forest on the mountain's western flank.

Mt. Ephraim Road emerges from the woods on Stronghold's northwestern edge at Bennett Creek. A short distance downstream you find the Lilypons Water Gardens, a commercial nursery and retail outlet for the aquatic gardening trade. The Lilypons property encompasses a broad area of low floodplain adjacent to the confluence of Bennett Creek and the Monocacy River. For years the owners have welcomed birders and general nature enthusiasts onto their property, whose habitat is a delightful mix of fields and ponds, some filled with water, others mostly mud flats, and yet others dry and weedy. This mixture of habitats supports a great variety of wildlife, with birds and frogs especially conspicuous. Lilypons Water Gardens is a great place to visit in early spring, when you might find a common loon or a tundra swan on one of the ponds, watch dozens of hawks overhead, and at dusk listen to a deafening chorus of frogs. Visit in summer to see both wild and cultivated aquatic plants in bloom. At any season watch for loggerhead shrikes, a rare bird in the mid-Atlantic; a pair has resided at Lilypons for many years.

Seneca Creek is a tributary of the Potomac River that flows through the rapidly developing suburban areas of Montgomery County, Maryland, yet a few natural areas along the creek and its main tributary, Little Seneca Creek, are fine places for nature study. Seneca Creek State Park protects several scattered areas along the creek, though most people only know the park for its developed heart off Clopper Road, near Gaithersburg. A tributary of the creek is dammed here to form Clopper Lake, a popular spot for fishing and turtle watching. Trails lead through woods and across an interesting power-line cut in this section of the park. Areas underneath power lines and along underground utility pipelines are generally kept clear of forest. The section of power-line clearing in the park makes a marvelous walk because its edges are filled with shrubby mountain laurels that burst into bloom during late spring, and the open areas fill with flowers and butterflies all summer long. The little wood satyr, a small brown butterfly with big circular spots on each wing, is more common in the upland sections of Seneca Creek State Park than at any other place I've seen. Watch for little wood satyrs between mid-May and mid-July. On the hilltops you can see some surprisingly nice vistas of the rolling Piedmont countryside. Trails that lead through the clearing also drop into the deep dark woods that line the stream, which is called Great Seneca Creek in this section.

A larger lake has formed behind a dam built on Little Seneca Creek. Most of the land surrounding Little Seneca Lake is managed as a local park called Black Hill. The dam was built in the early 1980s, and every naturalist that I know was surprised when almost immediately the new lake started attracting an impressive variety of waterfowl. I've been observing the lake formally once every year since its inception as part of the Sugarloaf Christmas Bird Count, and the list of waterbirds that I've seen here is extensive. Tundra swans visit the lake every winter, and a trumpeter swan was once found here. Ross' goose is another rarity that was once found on the lake. Usually during the colder months at least a dozen species of ducks can be found on the waters, and sometimes loons and grebes are also here. Land birding at Black Hill is also excellent, with a plentiful assortment of raptors and songbirds usually present at all seasons. Red-headed woodpeckers, which are rare in our region, have nested around the lake for several years.

My favorite section of the Black Hill Regional Park is the quiet northern corner, where Ten Mile Creek enters the Little Seneca Lake. A set of short trails leads along the lakeshore, around hilly fields, through rich, fairly mature deciduous forest, and past several small wetlands. Wood ducks and beavers can often be seen in this corner of the lake, ospreys nest here, and brushy hedgerows along the field edges are home to large numbers of chipmunks, cardinals, eastern towhees, and a variety of sparrows. During the warmer months this area is busy with butterflies and other insects; down by the lakeshore the dragonflies perform their dazzling acrobatic flight maneuvers. Ten Mile Creek must be crossed to reach these trails, but usually some rock hopping can get you dryly across. One thing's for sure; you'll never have to worry about crowds in this scenic and quiet section of Black Hill Park.

Most of the lower stretch of Seneca Creek is protected as an undeveloped section of the state park. Berryville Road touches the creek at one of its broad bends; follow the valley downriver from here (there is no formal trail, but travel isn't too difficult) and you'll see high bluffs surrounding deep forest groves where wild turkeys still roam. During times of moderately high water, especially in spring, I recommend a canoe trip along the lower section of Seneca Creek; wildflower displays in the creek valley are rich and varied.

Seneca Creek reaches the Potomac River just below the little community called Seneca. The town's major landmark is Poole's Store, which maintains the ambiance of the classic general store of the early twentieth century. Stop in for a snack or for the nostalgia. Follow the C & O Canal towpath upriver from the mouth of Seneca Creek to reach a wide section of the canal, the Seneca turning basin. Painted turtles, prothonotary warblers, and green herons can be found here during the warmer months. Behind the basin, the river's floodplain is bounded by a high bluff of deep red sandstone. This durable and beautiful stone was quarried at this site and transported along the C & O Canal for various building purposes. The well-known original Smithsonian building on the Mall in Washington, D.C. (often called *The Castle*) is built from Seneca sandstone, as were many locks and lockhouses along the canal. The sandstone blocks were cut at a mill built of the same rock. The walls and foundation of the mill still stand. Red-shouldered hawks and Louisiana waterthrushes nest in the wet woods that surround the old mill, and American toads are common here.

An old, low dam blocks the Potomac just below its confluence with Seneca Creek. The dam was originally built to feed water into the C & O

Canal, but it also backs up the waters of the Potomac and the lower reaches of Seneca Creek to form a slackwater section that is almost a lake. For years this area has been popular for fishing and for birding. Surprises I've found on the water here include red-necked grebe and black tern. This section of river has also recently become popular for recreational boating. On any nice day, especially weekends in spring, summer, or fall, Seneca seems overrun with large, high-powered speedboats and small but extremely noisy jet skis. Waterbirds are chased away, fishing boats are constantly disturbed, and the incredible noise makes it hard to even hear the songbirds. Once the confluence of Seneca Creek and the Potomac River was among my favorite spots but my visits are rather infrequent now.

ROCK CREEK

Georgetown is Washington D.C.'s oldest community. Long before a federal city was ever conceived back in colonial days, a town was founded near the upper reaches of the navigable portion of the Potomac River. A site was chosen where a stream flowed into the river, to take advantage of water power from the stream and a small safe harbor at the stream's mouth. That stream was Rock Creek.

In the early nineteenth century, Rock Creek was selected as the starting point for the Chesapeake and Ohio Canal. Barges would travel a short stretch of the creek between the Potomac River and the canal. A little farther upstream, near the Fall Line, were mills and rapidly growing communities. Soon city completely encircled the lower section of Rock Creek, and today it is clearly part of Washington, D.C.'s urban setting. In 1890, however, large areas of land in the northern part of Washington, D.C. surrounding Rock Creek were set aside as parkland, along with a narrow section bounding the creek farther south, right down to its junction with the Potomac. Today Rock Creek Park remains as a green oasis in the middle of the urban environment.

Rock Creek Park, managed by the National Park Service, is a surprisingly rich natural area. One section of the park has undoubtedly the best wildlife-viewing opportunities in our entire region, the National Zoo. Where else can you see a panda, giraffe, or a flock of flamingos? In addition to its wondrous exotic wildlife, a variety of native wildlife can be seen in and around the zoo, since the facility is tucked into the park. Wild wood ducks join captive waterfowl on the ponds in front of the main aviary, and black-crowned night-herons nest above the bald eagle cage.

From the zoo north to the Maryland line, Rock Creek Park widens and its forest becomes relatively wild and mature. Trails wind through various sections of the park past some impressively large tuliptrees, oaks, maples, beeches, and, down by the creek itself, sycamores. Spring wildflower diversity is excellent throughout the park's forests. Many of the park's open areas were converted from lawn to meadow back in the 1980s, and many summer-blooming wildflowers thrive in this habitat. Songbirds still nest in the park, though their diversity has been steadily declining. Still, certain areas remain outstanding during migration. The hilltop area surrounding the park's nature center, just south of Military Road, is particularly rich during early May, when twenty or more warbler species may be seen or heard on a good day. Another outstanding area is just south of the Maryland line, along the trail that begins near Boundary Bridge and follows the creek's east side. A pair of great horned owls always nests somewhere near this trail, and box turtles are frequently seen. Muddy spots along this trail are often filled with mammal tracks. White-tailed deer, opossum, and raccoon are common here and throughout Rock Creek Park.

Box turtle

In Maryland the Rock Creek Valley is managed as a local park, and pleasant forest lines the creek for most of its length, right up to its headwaters in north-central Montgomery County. For much of its Maryland length the park features a paved recreation path, often full of walkers, joggers, in-line skaters, and bicyclists. Actually, serious cyclists find the trail poorly designed for bicycling, so they generally follow the paved roads that run through the park, either Jones Mill Road or Beach Drive. By whatever means you choose to travel this section of the park, you will find many plants and animals to enjoy. Along the Jones Mill Road section of the park, just south of the Capital Beltway, are several small marshes and vernal pools that are havens for amphibians. Vernal pools are ephemeral, filling with water for part of the year (usually in winter and spring) and later completely drying out. Spring peepers, wood frogs, and spotted salamanders turn these wetlands into seething masses of amphibian activity in late March and early April; drive past these ponds at night and you'll hear the peepers calling even if your windows are tightly rolled up.

Just west of the park in this section is Woodend, formerly a private estate that is now the headquarters of the Audubon Naturalist Society. I've been visiting Woodend for twenty years, with great regularity over the last twelve since I was hired by the Audubon Naturalist Society. It's a great little nature sanctuary, especially considering its location inside the Capital Beltway. An extremely active nature education program for children and adults is based here, with hundreds of classes offered annually. The society has long been an important conservation voice for Maryland, D.C., and Virginia, beginning in 1897 when protection of wading birds was its first mission. Today's conservation agenda is far more complex and diverse, and the Audubon Naturalist Society is active in efforts to improve air quality, water quality, and wildlife habitat. The society publishes the monthly *Audubon Naturalist News,* which features a wide variety of conservation and natural history articles. Woodend also has a fine natural history bookstore. It's a wonderful place to go to work every morning!

The 40 acres (16 hectares) of Woodend are managed as a nature reserve, and short nature trails lead through the property. A surprising variety of wildlife has been sighted at Woodend over the years, with white-tailed deer and red fox now regularly seen. Birding can be good in any season at Woodend, with almost all of the region's migrant songbirds recorded here. Birders were astounded one winter when white-winged crossbills were seen feeding on spruce cones at the sanctuary's western edge; a few years later red cross-bills were found at the same place. Eastern screech-owls and indigo buntings nest here. A small pond teems with the activity of frogs, dragonflies, water beetles, and salamanders. Although most of the sanctuary is wooded, four meadows provide habitat for summer wildflowers, woodchucks, and butter-flies. The Audubon Naturalist Society has long avoided the use of pesticides on the sanctuary, and as a result its insect diversity is extremely high. And even though praying mantids, ladybugs, sphinx moths, tiger swallowtails, and many other interesting and beneficial insects may be observed at Woodend, troublesome mosquitoes and biting flies are quite uncommon. Fireflies are so abundant that David Attenborough came to Woodend when he wanted to film these luminescent beetles for a television program. We were a bit disap-pointed when he identified the site only as eastern North America on his program. The diversity of Woodend's wildlife is partly due to its connection with the extensive forests, wetlands, and meadows of Rock Creek Park. Other nearby habitat has disappeared as two smaller estates adjacent to Woodend have been subdivided, making the sanctuary even more important as an oasis for wildlife in the midst of the urban-suburban desert.

Good wildlife habitat lines the Rock Creek valley north to its headwaters northeast of Gaithersburg. The most extensive natural area in the watershed is north of Rockville at the Rock Creek Regional Park. The park features the Meadowside Nature Center, the Smith Center (Montgomery County's main environmental education facility), forests, fields, wetlands, and two small reservoirs. Lake Needwood lies along Rock Creek, and although a lot of recreational activity occurs on and around Needwood, it's still decent wildlife habitat, some autumns collecting a good variety of waterfowl. Lake Frank is on the North Branch of Rock Creek, and the only development here is a walking trail around the lake. The brushy habitat around Lake Frank supports a diversity of wildlife, from nesting prairie warblers to black rat snakes and five-lined skinks. Gardens next to the nature center and open meadows to the north are visited by many butterflies and other insects.

ANACOSTIA

The Anacostia is sometimes called Washington D.C.'s forgotten river. Like Rock Creek the Anacostia flows from north to south through Washington D.C.'s suburbs before winding through the city and emptying into the

Potomac. Although the Anacostia is just a few miles east of Rock Creek, it has a completely different character. For most of its length within the boundaries of Washington D.C., the Anacostia is a tidal river, its flow slow and changing direction with the tides. The broad, deep nature of a tidal river makes it navigable for a wide array of ships, and these characteristics influenced the development of the Anacostia. Whereas Rock Creek supported a few mills but little other construction, the Anacostia was used as a river of commerce and industry, and coincidentally as a dumping ground. The tidal Anacostia became polluted, its shores mostly developed, and its wetlands almost completely destroyed.

Fortunately a few areas were protected, and efforts have begun to improve the Anacostia environment. Water quality is slowly improving, and wetlands have been restored in several places. Two of the more interesting natural areas that have been protected in the Anacostia Valley lie across the river from each other near Washington, D.C.'s eastern corner. On the west bank of the river is the National Arboretum, where extensive horticultural plantings are bordered by protected natural habitats. Kenilworth Aquatic Garden, a national park facility, is on the east bank. Like the National Arboretum, Kenilworth has a pleasant mix of tended garden and protected natural area. Anacostia wetland habitat restoration projects have been centered along this section of the river. Both areas are good places for observing plants, birds, insects, and, especially at Kenilworth, frogs.

Upstream at Bladensburg, Maryland, the Anacostia's two main forks converge, known simply as the Northwest Branch and the Northeast Branch. Fine little natural areas may be found along both branches and their tributaries. Greenbelt Park is a small unit of the national park system along the Northeast Branch. Largely dry upland forest, Greenbelt Park offers forest trails that feature an excellent diversity of woody plants, shrubs being especially well-represented. Mountain laurel, blueberry, wild azalea, and several species of viburnum are common beneath the canopy of Virginia pine, sweetgum, red maple, and several kinds of oak.

The upper sections of Beaverdam Creek, which flows into the Northeast Branch, flow through the Beltsville Agricultural Research Center. Several roads lead through the center's mixture of research plots, open meadows, and farms, and although access on foot is limited, much interesting wildlife may be observed from the roads. Some of the earliest efforts to restore populations of eastern bluebirds occurred at Beltsville, and this lovely little bird is now quite common here. Another tributary of the Northeast Branch is Paint Branch, whose headwaters feature some of the suburban region's best water quality, good enough to support a population of brown trout. Quiet trails lead through the intact forest of the upper Paint Branch Valley. Downstream, in College Park, paved trails are more heavily traveled but still lead through protected forests and meadows.

The Northwest Branch has long been a favorite of mine because I grew up just a couple of miles from this stream. This branch flows past the flower-filled meadows of the National Capital Trolley Museum and through the woods of Wheaton Regional Park, sites of some of my earliest nature walks. From here to Colesville Road the Northwest Branch flows through an isolated wooded gorge, its hillsides supporting a fairly pristine, dry acid forest. Trailing arbutus, a relatively uncommon little wildflower in our region, is quite common in this gorge, its fragrant pink blossoms opening each April. Next to Colesville Road is a small dam that backs up the flow of the Northwest Branch for about half a mile. Beavers are active along this stretch, which is a great place for ice skating in cold winters. From here to Bladensburg the width of protected forest along the creek varies, yet enough remains to support a good assortment of plants and animals.

PATUXENT RIVER

The Patuxent River must be considered a suburban river. Its headwaters lie in the northern sections of Montgomery and Howard Counties, Maryland, and the river flows from these rapidly developing areas almost exactly halfway between Baltimore and Washington, D.C. on its way to the Chesapeake Bay. It's only about 80 miles as the crow flies from the Patuxent's source to its mouth, yet the river provides drinking water for a significant portion of the region, receives treated sewage and contaminated runoff in return, and still flows through a corridor that includes several parks and important wildlife habitat. In an effort to protect the river and its habitat, the Patuxent has been declared a Maryland Scenic River, but continued development within its watershed exacerbates problems associated with the quality of the water that flows into the river. Still, the river shows vitality, and its parks provide great opportunities for nature study right in the heart of the Baltimore-Washington metropolis.

A good chunk of the Patuxent's upper reach is protected as a part of

Patuxent River State Park, which also includes disconnected sections of the river valley downstream. Accessed from Georgia Avenue (Maryland Route 97), trails lead along the river's floodplain and into the surrounding uplands. Here you'll find the classic Piedmont forest dominated by oak and hickory trees. Rotting chestnut logs can still be found on the ground as reminders that the American chestnut tree was formerly codominant in this forest community. A few little wetlands are found in this section, great places to watch for bullfrogs, pickerel frogs, two-lined salamanders, and painted turtles.

An informal trail leads downstream from Georgia Avenue to the upper section of the Triadelphia Reservoir, the first of two small lakes created by dams on the Patuxent. As water-supply and flood-control facilities, the water level of each lake varies significantly—this section of Triadelphia can be quiet water or massive mud flat. This area seems to be rarely visited, though I know a few birders who have found migrant shorebirds feeding here. Access to the lake and its shores is restricted because the area protects a public water supply. Still, some boating and fishing are permitted at a handful of access points. Brighton Dam creates the reservoir, and the picnic area near the dam is a great spot for watching Baltimore orioles, which nest in the big shade trees. A fantastic azalea garden near the dam is open to the public only during the spring blooming period.

Below Brighton Dam the Patuxent winds and tumbles through mature floodplain forest before slackwater is again reached. This is the upper section of the second reservoir, named the T. Howard Duckett Reservoir but known to everyone locally as Rocky Gorge. Like the Triadelphia Reservoir, access to Rocky Gorge and the surrounding upland forest is limited to protect water quality. I won't say a word about the quiet little cove where friends and I used to go swimming years ago, or about the wild morels we gathered in the lakeside woods each spring.

Below Rocky Gorge the

Morel

Patuxent flows along the edge of Laurel, part of a rapidly growing region where towns and suburbs have almost completely replaced the farmland that separated Washington, D.C. from Baltimore just 30 years ago. Two tributaries, the Little Patuxent and the Middle Patuxent, flank the planned community of Columbia, Laurel's rival as the largest town between the two cities. Many small parks preserve wildlife habitat along these two branches; one of the planning objectives for Columbia was the protection of open space throughout the town. The main Patuxent below Laurel leads through a rich floodplain forest and wetland complex and then into an extensive protected area, the Patuxent Wildlife Research Center. As a primary research facility of the Biological Resources Division in the U.S. Geological Survey, the Patuxent center features projects that include the captive breeding of two endangered species, whooping crane and bald eagle. Most of the center is closed to public access and thus provides a haven for wildlife species that are sensitive to human activity. We have a tendency to love our urban and suburban parks too much, and too many visitors, even those sensitive to nature, result in the deterioration of the habitat's value to wildlife, especially sensitive species. I'd love to have free access to all of the research center's lands, but I'm glad that we've got some areas set aside primarily for the wildlife. The Patuxent center recently expanded when land that had been part of the Army's Fort Meade was transferred to the center.

The Patuxent River next winds between Bowie and Crofton, two other rapidly developing communities, before entering a still largely rural section between Bowie and Upper Marlboro. The farmlands in this stretch are attractive to open country wildlife, but runoff of sediments and agricultural chemicals from farms are a significant source of water pollution in the Patuxent itself. The Chesapeake Bay Foundation, a terrific organization dedicated to the protection of the Chesapeake Bay, runs a demonstration project called Claggett Farm where agricultural techniques that minimize water pollution may be observed. In fairness to farmers, it is worth noting that runoff of chemicals from suburban areas is also harmful to water quality and may have as great or greater cumulative negative effects. To protect the health of the Patuxent and of the entire Chesapeake, it is important to improve land use in both suburban and agricultural areas.

Below Upper Marlboro the Patuxent reaches that short enchanted stretch where it is tidal yet fresh. In our region most rivers have a stretch at the extreme upper limit of their tidal section where the flow of fresh water is great enough to maintain a freshwater habitat even though a tidal flow pushes in from the sea. On the Patuxent, this stretch occurs at a scenic spot known as Jug Bay. Happily, parks protect the river and its extensive freshwater wetlands on both banks here. On the west bank is Patuxent River Park, run as a Prince George's County Park by the Maryland National Capital Park and Planning Commission, and on the east bank is the Jug Bay Wetlands Sanctuary, an Anne Arundel County facility. Patuxent River Park features trails through an extensive swamp forest and along the river's shore. There's a great observation tower at the water's edge that offers a spectacular view of the broad Patuxent River and the marshes that line its shores. There is also access to the river, and this section's waters are ideal for traversing quietly by canoe. The wetlands harbor a diverse plant community, with wildrice dominant in the marsh. Showy flowering wetland plants at Jug Bay include buttonbush, cardinal flower, pickerelweed, and swamp milkweed. Birding is good year-round at Jug Bay, with late summer and fall an especially fine birding time because of the fruiting of wildrice. Many birds feed on the rice grains, including rails. The diminutive sora is common at Jug Bay from August through October.

The marshes and surrounding uplands are outstanding areas for a great variety of birds, reptiles, amphibians, butterflies, and other wildlife. Much is known about the diversity of this region thanks to the work of the Jug Bay Wetlands Sanctuary, which is managed more as a research center than as a park. Public access to this sanctuary is limited to certain days and specific events, but volunteers visit regularly to perform varied biological surveys. Their work has produced detailed resource inventories that elegantly illustrate the richness of this ecosystem. The sanctuary is a delight to explore, its varied trails leading through field and forest, to a boardwalk at the marsh's edge, and to a long causeway through the marsh; the causeway once was the grade for a railroad.

Downriver from Jug Bay the Patuxent gradually becomes brackish, its flow reversing twice daily with the tides and its shoreline meandering and often lined with brackish marsh. Big open fields next to the broad Patuxent at the Merkle Wildlife Sanctuary are a great spot for viewing eastern bluebirds and other open-country birds. Canada geese winter here by the thousands, and osprey are common nesters from here to the river's mouth. Other wetlands in the lower Patuxent are best explored by boat.

Eventually the broad, tidal Patuxent opens into the Chesapeake Bay at

Solomons Island. Along the way are several small tributaries; one of the most interesting is Battle Creek. Above the broad tidal stretch of Battle Creek lies a small baldcypress swamp, the only such swamp in Maryland west of the Chesapeake. There's an informative visitor center here along with a boardwalk trail that leads past the big swamp trees, which include baldcypress, red maple, and sweetgum. Take your time at this quiet spot and you might find snakes, turtles, or frogs. You'll surely hear birds, whose songs ring across the swamp's natural amphitheater.

PATAPSCO RIVER

Another relatively short yet geographically important river is the Patapsco. Its headwaters lie northwest of Baltimore in increasingly suburban Carroll County, Maryland, and it flows into the Chesapeake Bay in the industrial heart of downtown Baltimore. Like the Patuxent, the Patapsco must cope with the multiple indignities faced by a river in our rapidly developing metropolitan corridor. Yet parks protect significant portions of the Patapsco River Valley, providing excellent opportunities for nature study close to the homes of millions of people.

The Patapsco begins as two rivers. The South Branch flows through primarily rural Piedmont countryside along the Carroll-Howard County line, its rich floodplain forest accessible at many road crossings. A significant portion of the North Branch Valley is flooded under the Liberty Reservoir, part of Baltimore's municipal water supply, and a popular fishing spot. Several hiking trails lead through the varied forests that surround the reservoir, fine areas to visit in spring and summer to search for wildflowers and songbirds. Migrant shorebirds sometimes spend time on Liberty Reservoir's mud flats in late summer and fall. The branches meet just south of Liberty Reservoir in an area preserved as the McKeldin section of Patapsco Valley State Park. Here the river tumbles across several rapids surrounded by mature floodplain forest. Watch the river for signs of raccoons, beavers, and river otters. Climb the steep trails up from the river into oak forests where red-eyed vireos, eastern wood-pewees, and worm-eating warblers nest.

Just east of Liberty Reservoir is one of the most fascinating parks in our entire region, the Soldiers Delight Natural Environment Area. This is one of the few places in the mid-Atlantic where serpentine, a metamorphosed igneous rock, occurs at the surface. This uncommon rock erodes into a nutrient-poor soil that's loaded with metallic elements, including nickel, magnesium, and chromium. The presence of these elements suppresses the growth of many plants, leaving a niche for species with specialized adaptations for this environment. Fringed gentian, fameflower, and serpentine aster are among the uncommon species found in serpentine barren habitats. In places the habitat is prairielike, complete with prairie plants such as bluestem grass and blazing star. Blazing stars bloom in September, when they are an important nectar source for a variety of butterflies, including the uncommon leonard's skipper.

A tributary of the Patapsco's South Branch is Piney Run, and a Carroll County Park named for this creek protects about 500 acres (200 hectares) of forest in this watershed. Though largely managed for recreation, with a 300-acre (120 hectares) lake popular for fishing and an extensive picnic area, Piney Run has an excellent nature center and several short trails that run through oak woods, brushy fields, and along the lakeshore. Large flocks of waterfowl sometimes gather in this lake during fall, with the rare Eurasian wigeon recorded here at least once.

The Patapsco gathers momentum beyond the convergence of its two branches, and its valley broadens as the river heads toward its urban ending, passing along the edges of Ellicott City and Catonsville before entering Baltimore. Fortunately most of this section is protected as Patapsco Valley State Park. There are many access points to the park, whose forests are at their best in spring when wildflowers bloom and migrant songbirds mix with the nesting species. I'm especially fond of the area known as Hollofield, just south of U.S. 40, where scenic views over the forested valley show almost no sign of human development, even though downtown Baltimore is less than a dozen miles away. The section of the park farther downstream, accessed from Hilton Avenue, is even closer to the city, yet it remains surprisingly wild and peaceful.

GUNPOWDER FALLS

First a definition: Gunpowder Falls is the name of an entire river, not just a feature of the river. The river's main tributary is known as Little Gunpowder Falls. Once Gunpowder Falls reaches sea level it becomes an arm of the Chesapeake Bay, and here it is called the Gunpowder River. This is another of our clever plots to confuse newcomers to the area with bizarre place names.

Like the Patuxent and the Patapsco, the Gunpowder is another fairly short river whose headwaters lie in the Piedmont, whose waters are dammed for municipal water supply, and whose wooded valley is protected in many places as parkland. I think of the Gunpowder and the Patapsco as bookends that surround Baltimore, the Gunpowder being a little wilder because it's a little farther away from the city. The river is dammed twice, forming the Prettyboy Reservoir and the Loch Raven Reservoir, both in Baltimore County, Maryland. Each lake is surrounded by forest reserves whose primary purpose is to protect water quality. These forests effectively protect wildlife too, and trails lead through the oak, hickory, and tuliptree forest that surrounds each reservoir. Loch Raven, because it is closer to the city, has more visitors, many who come to fish. Still, the quiet corners of Loch Raven sometimes harbor an interesting mix of waterfowl during the colder months, especially in autumn.

Gunpowder Falls State Park protects several widely scattered sections of the river's valley. The Hereford section of the park lies just below the Prettyboy Reservoir, and features trails that lead along the river's floodplain, where displays of Virginia bluebells and other April wildflowers are an annual highlight. Other sections of the park protect forests along the Little Gunpowder Falls and wetlands at the river's junction with the Chesapeake. The Hammerman section of the park, located at the river's mouth, is heavily developed for recreation, yet a freshwater tidal marsh and some field and forest habitat are protected here. The Days Cove section of Gunpowder Falls State Park is across the mouth of a tributary called Bird River. Here are wetlands to explore without the crowds.

For many years the North Central Railroad followed the valley of Gunpowder Falls and the tributaries of Little Falls and Beetree Run along a north-south transect from Pennsylvania to Baltimore. The trains and the tracks are now gone, but the right-of-way has been acquired as an addition to Gunpowder Falls State Park and is now maintained as a hiking or bicycling trail. Tucked down in the floodplain forest, the trail is popular on weekends when the weather is nice, yet it always feels somewhat secluded. The forest is broken here and there by scraggly meadows, fine spots to pick blackberries during early summer and to watch for butterflies from spring through fall. Indigo buntings seem to nest at the edge of each break in the forest.

OCCOQUAN RIVER

One of the most significant tributaries of the Potomac south and west of Washington, D.C. is the Occoquan River, which empties into the tidal

Potomac between the town of Woodbridge and the Mason Neck peninsula. Its watershed includes many small creeks in Prince William, Loudoun, and Fairfax Counties, Virginia. One of the largest of these creeks is Bull Run, which gathers some of its waters from a low ridge of erosion-resistant greenstone that divides Virginia's Piedmont, appropriately named Bull Run Mountain.

For much of its length Bull Run is a classic little Piedmont creek, alternating between small rapids and quiet pools. It winds its way through pastoral farm country and then into floodplain forest where sycamores and silver maples tower above showy displays of Virginia bluebells, spring beauties, blue violets, yellow violets, and other wildflowers each April. The floral display is especially rich along the creek near the rapidly growing town of Manassas, a section preserved as Bull Run Regional Park. The Bull Run-Occoquan Trail begins in the park and follows the Bull Run Valley downstream for more than 17 miles. For its entire length this trail leads through excellent wildlife habitat, mostly floodplain forest. The trail crosses Hemlock Overlook Regional Park, where a dense grove of hemlocks includes some trees more than 200 years old.

Another part of the watershed that won't be developed is the Manassas National Battlefield Park, site of two major Civil War battles. For historical reasons many such battlefields in Virginia are protected from development, and each doubles as an interesting natural history destination. At Manassas forest and field habitats are preserved, making this park a particularly good site for observing butterflies and wildflowers in summer and for watching red-tailed and red-shouldered hawks in winter. Eastern bluebirds, grasshopper sparrows, and field sparrows are among the birds that nest in the meadows of the battlefield. Several native grasses, including little bluestem and Indian grass, grow in the meadows, along with summer-blooming asters, goldenrods, and less conspicuous wildflowers such as wild petunia and green milkweed.

As is true of most of the rivers close to our cities, the Occoquan has its dams, with several small reservoirs on its tributaries and a large lake, the Occoquan Reservoir, extending nearly 10 miles from just east of Manassas almost all the way to the Potomac. Several access points to the lake attract those interested in fishing and boating, and a few trails lead through the forest surrounding the lake. Fountainhead Regional Park provides access to the Occoquan Reservoir. This park is managed more for recreation than for nature, though its forests, fields, and lakeshore habitats support a varied wildlife community. The Bull Run-Occoquan Trail's downstream terminus is located here.

Below the dam and the town of Woodbridge, at the southern edge of the river's mouth, is a tract of land that for many years was a military reserve known as the Harry Diamond Lab. Downsizing of the military made this land available to buy, and developmental interests tried to purchase the land from the federal government. As is true for many military lands across the country, though, security buffers had inadvertently protected outstanding natural areas. Conservation groups successfully lobbied for protection of this property, which was designated an addition to the Mason Neck National Wildlife Refuge. Public access to this tract is now restricted, though it may someday be allowed. It's worth noting that the military owns several other large tracts of land in northern Virginia, including Fort Belvoir and the Quantico Marine Corps facility. If the military gives up any of this land in future years, let's hope that the conservation community will once again mobilize and secure protection for the forests and wetlands found on these properties.

HUNTLEY MEADOWS PARK

Huntley Meadows Park is a remarkably rich wetland park tucked into the middle of suburban Alexandria just a few miles southwest of Washington, D.C.'s southern tip. The heart of Huntley Meadows is a large freshwater marsh and swamp, which visitors can see up close from a boardwalk and observation tower. Beavers and muskrats are active in the marsh, and droppings of raccoons are often found on the

King rail

boardwalk. Birding at Huntley Meadows is extraordinary. For many years one birding highlight has been the park's nesting king rails. These marsh birds are extremely secretive in most places, but for some reason the king rails at Huntley Meadows have grown oblivious to humans, often walking right in front of people on the boardwalk. Other secretive marsh birds are also seen with surprising regularity at Huntley Meadows, including American bittern, least bittern, sora, and Virginia rail. With luck any of these may be seen during the warmer months of the year, as they slink between attractive marsh plants such as buttonbush and lizard's-tail.

Huntley Meadows is a fine spot to visit in winter, too. Throughout the colder months a variety of ducks join the resident Canada geese and great blue herons in the marsh. Many songbirds may be found at any season. Over-wintering birds such as cedar waxwing, eastern bluebird, and yellow-rumped warbler are joined by the first spring migrants—eastern phoebe, tree swallow, eastern towhee, and palm warbler—as early as March. Surrounding the wetland is a rich mix of young forest, mature forest, and old field. This combination of protected habitats supports many different plants and animals. Butterflies provide a good example of Huntley Meadows' diversity: Their populations were studied from 1978 to 1982 by butterfly expert Paul Opler, who found 79 species in and around the park during his 5-year study. Frogs, turtles, and water snakes are also abundant at Huntley Meadows and are most conspicuous during the warmer months.

The first time I visited Huntley Meadows I couldn't believe that such a rich and vibrant wetland habitat could exist so close to the city. I was hooked instantly, and I went back to the park regularly for the next year or so. I discovered a core of regular visitors, young and old, most of whom were excellent naturalists. These naturalists were generous in their tips about the park's habitats, its wildlife, and its seasonal cycles. Most lived nearby. There's a unique value to natural areas such as Huntley Meadows that are close to the homes of a large human population, for it is at these areas where many of us observe nature in a regular, detailed manner.

Suburban encroachment has threatened Huntley Meadows time after time, with proposals for new roads across the park the most frequent threat, but conservationists and park advocates have thus far kept new road development out of Huntley Meadows. A core group of the park's regular visitors organized as the Friends of Huntley Meadows, and this group can take much of the credit for maintaining the park's integrity through its efforts to influence decision makers and to mobilize the community. Here lies another important value to the local park: When we mobilize as a community to protect a favorite natural area, hopefully we raise community awareness about general environmental health and the need to protect natural areas nationally and globally. Every time I visit Huntley Meadows I delight in the richness of its varied habitats and give a quiet nod of thanks to those who have fought for its protection. Dyke Marsh, a small but biologically rich wetland on the Virginia shoreline of the Potomac just a few miles to the east, has similarly been protected by an active citizens' support group.

PRINCE WILLIAM FOREST PARK

South of the Occoquan and adjacent to the Quantico Marine Corps facility is Prince William Forest Park, a unit of the national park system. Like many of our mid-Atlantic parks, this is a restoration project. Two centuries of intensive agriculture in this area depleted the soil of much of its value. The abandoned land was gradually reclaimed by forest, first Virginia pines and red cedars, then oaks and other hardwoods. Prince William Forest Park is now mostly wooded, though the forest has a sparse look because of the dryness of the soil and its paucity of nutrients. Still, it's a fine place for forest walks, and wildlife diversity is relatively high. It's an excellent place to observe the ecological process known as succession, a process by which the land can often slowly recover from past abuses and bring back forest, certainly a hopeful sign in this era of accelerating environmental degradation.

LOCAL VIRGINIA PARKS AND NATURE CENTERS

We're fortunate throughout the Baltimore-Washington region to have many local parks and nature centers in the midst of our communities. This is true in northern Virginia, and although I won't discuss these areas in depth, I encourage residents of Virginia to note which local park is closest to home and to visit the park regularly to observe seasonal changes and transitions in their local environments. Nature centers close to Washington include Long Branch, Gulf Branch, Eleanor Lawrence, Riverbend, Hidden Oaks, Hidden Pond, the National Wildlife Federation's Laurel Ridge, and Potomac Overlook.

A few years back I bought a bicycle, the first I'd owned since childhood. Somehow, through college days and over subsequent years, when many friends rode sleek touring bicycles, I never made the purchase that allowed me to join them. Anyway, those bikes always looked so fragile, and I was frightened by the concept of riding on the roads with Washington drivers.

Several factors led to my decision to finally buy a bike. I worked for the Audubon Naturalist Society, an environmental organization, and lived fairly close to my workplace; to commute by bicycle would be an appropriately green activity. A new generation of bicycles had arrived, with fat tires and straight handlebars that seemed far less intimidating (and far sturdier) than those spindly touring models; these were bikes one could ride to places other than the scary roads. A good sale price caught my eye and instantly I became a cyclist.

Although I don't make the commute to work on my bicycle nearly as often as my environmental conscience thinks I should, much to my surprise I've discovered that bicycling can be a great way to explore the natural world. I've found a few places where trails or old roads are closed to cars but open to bikes. Riding slowly in such a place allows me to hear bird songs, spot wildflowers and butterflies, and choose promising habitats at which to stop, get off the bike, and explore. Some areas well-suited for bicycle travel have been discussed elsewhere: the C & O Canal along the Potomac River, the North Central Railroad Trail along Gunpowder Falls, Great Dismal Swamp in southeastern Virginia, and the Greenbrier River Trail in West Virginia. One of the best places for nature study from a bicycle is northern Virginia's Washington & Old Dominion Trail, known generally as the "W & OD."

The W & OD stretches from Alexandria to Purcellville, 45 miles to the west. This linear park follows the right-of-way of an abandoned railway whose name was transferred to the park. The rails have been replaced by a paved trail. There are open meadows, deciduous forests, gentle creeks, and other habitats to explore along this park corridor. The trail passes through quiet neighborhoods, past the small town heart of suburban communities such as Herndon and Vienna, next to golf courses, strip malls, and a gravel pit. There's even a way station with bathrooms, soda machines, and an airhose for inflating tires. In the mixed forest and open country lining the

trail many eastern bluebirds, prairie warblers, yellow-breasted chats, and indigo buntings nest. Butterflies love the flowery meadows and old fields, and
more than 450 species of plants have been catalogued along the trail. Many
of these have been observed at Clark's Crossing Park, which straddles the
trail near Vienna. A visit to the meadow here is especially rewarding in summer, when the open-country wildflowers bloom. Butterfly weed, swamp
milkweed, New York ironweed, asters, goldenrods, and other nectar-rich
plants attract an excellent variety of butterflies and other insect pollinators.
West of Herndon the scenery changes as the rolling Piedmont reaches out
toward the Blue Ridge. All sections of the W & OD can be explored on
foot, but I find it delightful to explore on a bicycle, partly because of its ideal
design as a cycling path.

New bicyclists learn a sad truth quickly, that most urban-suburban bike
paths are rotten places for bicycling. Paths are usually narrow with quirky
turns and frequent road crossings; these are great spots for walking, jogging,
exercising the dog, in-line skating, pushing the stroller, and so forth but bicyclists trying to ride steadily along these paths encounter a stream of obsta-
cles to navigate. Conflicts can be so frequent that hostilities emerge, as once
happened on the Mount Vernon Parkway's bike path when an angry noncy-
clist spread tire-popping tacks along the path.

The W & OD is a happy exception. Actually, for much of its length, a
pair of trails has been developed, one paved for wheeled travel and a parallel
dirt path for walkers, joggers, and equestrians. The path is wider than most
bike paths, regular signs describe the trail's protocol and rules, and, surprisingly important, there is a center line. Everyone seems to know how to react
when a voice from behind calls, "Passing on the left." The many users of this
park seem able to coexist marvelously, thanks to the path's design. There's
even a place for me and my bicycle among the trail's users. Just one warning
to my fellow W & OD cyclists: If you're behind me on the path, don't follow too closely. I'll be stopping soon, no doubt, to check out that bird, plant,
turtle, or butterfly. I'm not out to cover a great number of miles or to win
races. My bicycle is the latest tool I use for nature study, and I've learned that
it can be just as useful as binoculars and field guides.

As I close this chapter, I think about all the little places I haven't mentioned and am thankful we've managed to protect so many spots where nature can still be observed. I wish most parks were bigger, I wish the historical wildlife communities were intact, I wish that native plants weren't so

often battling aggressive introduced species such as Japanese honeysuckle and multiflora rose. I wish that just one big section of primeval mid-Atlantic deciduous forest ecosystem, maybe 5,000 acres, had been preserved. But I rejoice that we still have so many places for nature study, with more species of plants and animals than I'll ever be able to find. Part of our planet's great charm is its incredible diversity of life. It's a great comfort to know that, even in the midst of urban–suburban environmental degradation, nature is still too complex to ever completely understand and master. How horrid it would be to have nothing left to learn.

Epilogue

It's difficult to finish writing this book. I've left out many places that I know, and many more that I still want to visit for the first time. I've skipped the stories of many plants, animals, and geological features of the area, and my understanding of those that have been included is incomplete. The world is just too complex for me to completely understand and synthesize, even when I look at limited geographical areas. Although this complexity is a bit intimidating, it is also reassuring. I hope to never see the day when there are no new places to visit, no new discoveries to make, no unsolved mysteries in nature.

It's my hope that this book helped you develop some understanding of nature in the mid-Atlantic. I hope it also encourages many of you to head afield and explore, either at areas mentioned in the text or at some of the places I didn't describe. For those who wish to learn more, the following appendixes that list books and organizations may prove helpful.

Appendix 1.
A Beginning Naturalist's
Reference Library

This list of books should help you build a beginner's reference library on mid-Atlantic natural history. For complete bibliographic citation, please see References.

To identify wildflowers, I recommend *Newcomb's Wildflower Guide* (Newcomb 1977) or *A Field Guide to Wildflowers* (Peterson and McKenny 1968). For trees I prefer *Trees of North America* (Brockman, Zim, and Merrilees 1968), for ferns *A Field Guide to the Ferns* (Cobb 1963), and for grasses *Grasses, An Identification Guide* (Brown 1979). The more technical works on plants have the advantage of including all species known to occur within a stated region. For this reason, and because each is more user-friendly than many other technical works, I often use *Flora of West Virginia* (Strausbaugh and Core 1978), *Woody Plants of Maryland* (Brown and Brown 1972), and *Herbaceous Plants of Maryland* (Brown and Brown 1984). Fungi are well described in *Mushrooms of North America* (Miller 1980) and *A Field Guide to Mushrooms* (McKnight and McKnight 1987). To identify lichens I recommend *How to Know the Lichens* (Hale 1979).

There are many field guides to vertebrate animals. For birds, *Field Guide to the Birds of North America* (National Geographic Society 1983) is my first choice. If you choose another book, select one using artwork, not photographs. For reptiles and amphibians, two regional books are most useful and together cover all mid-Atlantic species: *Amphibians and Reptiles in West Virginia* (Green and Pauley 1987) and *Amphibians and Reptiles of the Carolinas and Virginia* (Martof et al. 1980). *Mammals of the Carolinas, Virginia, and Maryland* (Webster, Parnell, and Biggs 1985) is another fine regional reference. To identify freshwater fish, use *A Field Guide to Freshwater Fishes* (Page and Burr 1991); for coastal areas use either *Life in the Chesapeake Bay* (Lippson and Lippson 1984) or *Chesapeake Bay: Nature of the Estuary* (White 1989). Both of these last two books also include information about other animals and plants of the Chesapeake Bay region.

Begin a study of insects with *A Field Guide to Insects* (Borror and White 1970). For butterflies I recommend *A Field Guide to Eastern Butterflies* (Opler and Malikul 1992), *Butterflies East of the Great Plains* (Opler and Krizek 1984), or *Butterflies Through Binoculars* (Glassberg 1993). *A Field Guide to Moths* (Covell 1984) and *A Field Guide to Beetles* (White 1983) will add to your appreciation of insects. *Beachcomber's Guide from Cape Cod to Cape Hatteras* (Keatts 1995) and *A Field Guide to the Atlantic Seashore* (Gosner 1978) help identify coastal invertebrates.

For ideas about places to visit, begin with *Finding Birds in the National Capital Area* (Wilds 1983) and *Finding Wildflowers in the Washington-Baltimore Area* (Fleming, Lobstein, and Tufty 1995). Other useful books include *The Smithsonian Guides to Natural America: The Atlantic Coast and Blue Ridge* (Ross 1995), *Natural Washington* (Berman and Gerhard 1994), *The Audubon Society Field Guide to the Natural Places of the Mid-Atlantic States: Coastal* (Lawrence 1984), and *The Audubon Society Field Guide to the Natural Places of the Mid-Atlantic States: Inland* (Lawrence and Gross 1984).

Appendix 2.
Conservation Organizations, Nature Centers, Parks, Wildlife Refuges, and Related Resources

There are many organizations and facilities in the mid–Atlantic region whose properties, publications, and human resources can be of great help to outdoor enthusiasts and students of nature. Contact information for some of these resources is listed in this appendix. Remember that mailing addresses and phone numbers can change over time.

Algonkian Regional Park
1600 Potomac View Road
Sterling, VA 22170
703-450-4655

Antietam National Battlefield
Box 158
Sharpsburg, MD 21782
301-432-5124

Appalachian Trail Conference
Washington and Jackson Streets
P. O. Box 807
Harpers Ferry, WV 25425
304-535-6331

Assateague Island National Seashore
Box 294
Berlin, MD 21811
410-641-3030

Assateague State Park
Box 293
Berlin, MD 21811
410-641-2120

Audubon Naturalist Society
8940 Jones Mill Road
Chevy Chase, MD 20815
301-652-9188

Back Bay National Wildlife Refuge
4005 Sandpiper Road
P. O. Box 6286
Virginia Beach, VA 23456
804-721-2412

Baltimore Zoo
Druid Hill Park
Baltimore, MD 21217
410-396-7102

Battle Creek Cypress Swamp Sanctuary
c/o Calvert County Courthouse
Prince Frederick, MD 20678
410-535-5327

Beltsville Agricultural Research Center
Visitor Center
Building 302, BARC-East
Beltsville, MD 20705
301-344-2483

Black Hill Regional Park
20930 Lake Ridge Drive
Boyds, MD 20841
301-972-9396

Blackwater National Wildlife Refuge
2145 Key Wallace Drive
Cambridge, MD 21613
410-228-2677

Blue Ridge Parkway
200 BB&T Building
Asheville, NC 28801
704-298-0398

Bombay Hook National Wildlife Refuge
RFD 1, Box 147
Smyrna, DE 19977
302-653-9345

Brookside Nature Center
1400 Glenallen Avenue
Wheaton, MD 20902
301-946-9071

Bull Run Regional Park
7700 Bull Run Drive
Centreville, VA 22020
703-631-0550

Burke Lake Park
7315 Ox Road
Fairfax Station, VA 22039
703-323-6600

Cacapon State Park
Berkeley Springs, WV 25411
304-258-1022

Caledon Natural Area
11617 Caledon Road
King George, VA 22485
703-663-3861

Calvert Cliffs State Park
Route 4, Box 106A
Brandywine, MD 20613
301-888-1622

Calvert Marine Museum
P.O. Box 97
Solomons, MD 20688
410-326-2042

Canaan Valley State Park
Route 1-330
Davis, WV 26260
304-866-4121

Cape Henlopen State Park
42 Cape Henlopen Drive
Lewes, DE 19958
302-645-6852

Cape May Bird Observatory
P. O. Box 3
Cape May Point, NJ 08212
609-884-2736

Cape May Point State Park
P. O. Box 107
Cape May Point, NJ 08212
609-884-2159

Catoctin Mountain Park
National Park Service
6602 Foxville Road
Thurmont, MD 21788
301-663-9388

Cedarville State Forest
11704 Fenno Road
Upper Marlboro, MD 20772
301-888-1410

Chesapeake & Ohio Canal National
    Historical Park
Great Falls Tavern
11710 MacArthur Boulevard
Potomac, MD 20854
301-299-3613
and
Chesapeake & Ohio Canal National
    Historical Park
Box 158
Sharpsburg, MD 21782
301-432-5124

Chincoteague National Wildlife Refuge
P. O. Box 62
Chincoteague, VA 23336
804-336-6122

Clearwater Nature Center
11000 Thrift Road
Clinton, MD 20735
301-297-4575

Cranberry Mountain Visitor Center
Monongahela National Forest
Richwood, WV 26261
304-653-4826

Cunningham Falls State Park
14039 Catoctin Hollow Road
Thurmont, MD 21788
301-271-7574

Delmarva Ornithological Society
P.O. Box 4247
Greenville, DE 19807

Delaware Division of Fish and Wildlife
P.O. Box 1401
Dover, DE 19903
302-739-5297

Delaware Seashore State Park
Inlet 850
Rehoboth Beach, DE 19971
302-227-2800

Eastern Neck National Wildlife Refuge
1730 Eastern Neck Road
Rock Hall, MD 21661
410-639-7056

Eastern Shore of Virginia National
    Wildlife Refuge
RFD 1, Box 122B
Cape Charles, VA 23310
804-331-2760

Elk Neck Demonstration Forest
130 McKinnytown Road
North East, MD 21901
410-287-5675

Elk Neck State Park
4395 Turkey Point Road
North East, MD 21901
410-287-5333

Fairfax County Park Authority
3701 Pender Drive
Fairfax, VA 22030
703-246-5700

False Cape State Park
4001 Sandpiper Road
Virginia Beach, VA 23456
804-426-7128

Edwin B. Forsythe National
    Wildlife Refuge
Great Creek Road
P. O. Box 72
Oceanville, NJ 08231
609-652-1665

Fort Frederick State Park
11100 Fort Frederick Road
Big Pool, MD 21711
301-842-2155

Gambrill State Park
21843 National Pike
Boonsboro, MD 21713
301-791-4767

Gathland State Park
900 Arnoldstown Road
Jefferson, MD 21755
301-293-2420

George Washington Memorial Parkway
Turkey Run Park
McLean, VA 22101
703-285-2600

George Washington National Forest
Harrison Plaza
P. O. Box 233
Harrisonburg, VA 22801
703-433-2491

Great Dismal Swamp National Wildlife
    Refuge
P. O. Box 349
Suffolk, VA 23434
804-539-7479

Great Falls Park
P.O. Box 66
Great Falls, VA 22066
703-285-2966

Green Ridge State Forest
Star Route
Flintstone, MD 21530
301-478-2991

Green Spring Gardens Park
4603 Green Spring Road
Alexandria, VA 22312
703-642-5173

Greenbelt Park
6565 Greenbelt Road
Greenbelt, MD 20770
301-344-3948

Greenbrier State Park
Route 2, Box 235
Boonsboro, MD 21713
301-791-4767

Gulf Branch Nature Center
3608 North Military Road
Arlington, VA 22207
703-358-3403

Gunpowder Falls State Park
10815 Harford Road
P.O. Box 5032
Glen Arm, MD 21057
410-592-2897

Harpers Ferry National Historical Park
P.O. Box 65
Harpers Ferry, WV 25425
304-535-6224

Hawk Mountain Sanctuary
Route 2, Box 191
Kempton, PA 19529
215-756-6961

Hemlock Overlook Regional Park
13220 Yates Ford Road
Clifton, VA 22024
703-830-9252

Hidden Oaks Nature Center
4020 Hummer Road
Annandale, VA 22003
703-941-1065

Hidden Pond Park
8511 Greeley Boulevard
Springfield, VA 22152
703-451-9588

Horsehead Wetlands Center
P. O. Box 519
Grasonville, MD 21638
410-827-6694

Huntley Meadows Park
3701 Lockheed Boulevard
Alexandria, VA 22306
703-768-2525

Interstate Commission of the Potomac
    River Basin
Suite 300
6110 Executive Boulevard
Rockville, MD 20852

Lake Accotink Park
7500 Accotink Park Road
Springfield, VA 22150
703-569-3464

Lake Fairfax Park
1400 Lake Fairfax Park
Reston, VA 22090
703-471-5414

Laurel Ridge Conservation
    Education Center
National Wildlife Federation
8925 Leesburg Pike
Vienna, VA 22184
703-790-4439

Ellanor C. Lawrence Park
5040 Walney Road
Chantilly, VA 22021
703-631-0013

Leesylvania State Park
16236 Neabsco Road
Woodbridge, VA 22191
703-670-0372

Lilypons Water Gardens
P.O. Box 10
Buckeystown, MD 21717
301-874-5133

Little Bennett Regional Park
23701 Frederick Road
Clarksburg, MD 20871
301-972-6581

Locust Grove Nature Center
Cabin John Regional Park
7400 Tuckerman Lane
Rockville, MD 20852
301-299-1990

Long Branch Nature Center
625 South Carlin Springs Road
Arlington, VA 22204
703-358-6535

Manassas National Battlefield
6511 Sudley Road
Manassas, VA 22110
703-361-1339

Maryland Department of
    Natural Resources
580 Taylor Avenue
Annapolis, MD 21401
410-974-3195

Maryland Geological Survey
2300 St. Paul Street
Baltimore, MD 21213

Maryland Ornithological Society
Cylburn Mansion
4915 Greenspring Avenue
Baltimore, MD 21209

Maryland Ornithological Society
Montgomery County Chapter
P.O. Box 59639
Potomac, MD 20859

Maryland-National Capital Park and
    Planning Commission
9500 Brunett Avenue
Silver Spring, MD 20901
301-495-2525

Mason District Park
6621 Columbia Pike
Annandale, VA 22003
703-941-1730

Mason Neck National Wildlife Refuge
14416 Jefferson Davis Highway, Suite 20A
Woodbridge, VA 22191
703-690-1297

Mason Neck State Park
7301 High Point Road
Lorton, VA 22079
703-339-7265

Meadowside Nature Center
5100 Meadowside Lane
Rockville, MD 20853
301-924-4141

Merkle Wildlife Sanctuary
11704 Fenno Road
Upper Marlboro, MD 20772
301-888-1410

Monongahela National Forest
Elkins, WV 26241
304-636-1800

Morven Park
Route 3, Box 50
Leesburg, VA 22075
703-777-2414

Myrtle Grove Wildlife Management Area
Star Route Box 2209
La Plata, MD 20646
301-743-5161

National Capital Parks—East
1900 Anacostia Drive, SW
Washington, DC 20021
202-426-6905

National Capital Region
National Park Service
1100 Ohio Drive, SW
Washington, DC 20242

National Colonial Farm
3400 Bryan Point Road
Accokeek, MD 20607
301-283-2115

National Wildlife Visitor Center
10901 Scarlet Tanager Loop
Laurel, MD 20708
301-497-5760

National Zoological Park
Smithsonian Institution
Washington, DC 20008
202-673-4800 (recorded information)

The Nature Conservancy:
    Delaware Chapter
321 South State Street
Dover DE 19903
302-674-3550

The Nature Conservancy:
    Maryland Chapter
2 Wisconsin Circle, Suite 600
Chevy Chase, MD 20815
301-656-8673

The Nature Conservancy:
    Pennsylvania Chapter
1211 Chestnut Street
Philadelphia, PA 19107
215-963-1400

The Nature Conservancy:
    Virginia Chapter
1110 Rose Hill Drive
Charlottesville, VA 22901
804-295-6106

The Nature Conservancy:
    West Virginia Chapter
P.O. Box 3754
Charleston, WV 25337
304-345-4350

New Germany State Park
Route 2, Box 62A
Grantsville, MD 21536
301-895-5453

Northern Virginia Regional
   Park Authority
5400 Ox Road
Fairfax Station, VA 22039
703-352-5900

Nottingham Barrens
150 Park Road
Nottingham, PA 19362
215-932-9195

Oregon Ridge Park
13555 Beaver Dam Road
Cockeysville, MD 21030
410-887-1815

Patapsco Valley State Park
1100 Hilton Avenue
Baltimore, MD 21228
410-747-6602

Patuxent River Park
16000 Croom Airport Road
Upper Marlboro, MD 20772
301-627-6074

Patuxent River State Park
11950 Clopper Road
Gaithersburg, MD 20787
301-924-2127

Patuxent Wildlife Research Center
11400 American Holly Drive
Laurel, MD 20708
301-597-5592

Piney Run Park
30 Martz Road
Sykesville, MD 21784
410-795-3274

Pocomoke River State Forest and Park
3461 Worcester Highway
Snow Hill, MD 21863
410-632-2566

Point Lookout State Park
Route 5, Box 48
Scotland, MD 20687
301-872-5688

Potomac Appalachian Trail Club
118 Park Street, SE
Vienna, VA 22180
703-242-0315

Potomac Backpackers
P.O. Box 403
Merrifield, VA 22116
703-524-1185

Potomac Overlook Regional Park
2845 Marcey Road
Arlington, VA 22207
703-528-5406

Potomac Peddlars Touring Club
P.O. Box 23601
Washingtion, DC 20026
202-363-8687

Prime Hook National Wildlife Refuge
RD 3, Box 195
Milton, DE 19968
302-684-8419

Prince William Forest Park
Box 209
Triangle, VA 22172
703-221-7181

Raptor Society of
   Metropolitan Washington
P.O. Box 482
Annandale, VA 22003

Riverbend Nature Center
8814 Jeffery Road
Great Falls, VA 22066
703-759-3211

Rock Creek Nature Center
5200 Glover Road, NW
Washington, DC 20015
202-426-6829

Rock Creek Regional Park
6700 Needwood Road
Derwood, MD 20855
301-948-5053

Rocks State Park
3318 Rocks Chrome Hill Road
Jarrettsville, MD 21084
410-557-7994

Rocky Gap State Park
Route 1, Box 90
Flintstone, MD 21530
301-777-2138

Sandy Point State Park
800 Revell Highway
Annapolis, MD 21401
410-757-1841

Savage River State Forest
349 Headquarters Lane
Grantsville, MD 21536
301-895-5759

Seashore State Park and Natural Area
2500 Shore Drive
Virginia Beach, VA 23451
804-481-2131

Seneca Creek State Park
11950 Clopper Road
Gaithersburg, MD 20878
301-924-2127

Shenandoah National Park
Route 4, Box 348
Luray, VA 22835
703-999-2266

Sky Meadows State Park
Route 1, Box 540
Delaplane, VA 22025
703-592-3556

Smithsonian Environmental
  Research Center
P.O. Box 28
Edgewater, MD 21037
301-261-4190

Smithsonian Institution
National Museum of Natural History
Washington, DC 20560
202-357-2700

Soldiers Delight Natural
  Environment Area
5100 Deer Park Road
Owings Mills, MD 21117
410-922-3044

Sugarloaf Mountain
7901 Comus Road
Dickerson, MD 20842
301-869-7846

Susquehanna State Park
801 Stafford Road
Havre de Grace, MD 21078
410-939-0643

Swallow Falls State Park
RFD 5, Box 122
Oakland, MD 21550
301-334-9180

U.S. Geological Survey
2201 Sunrise Valley Drive
Reston, VA 22092

U.S. National Arboretum
3501 New York Avenue, NW
Washington, D.C. 20002
202-475-4815

Virginia Department of Conservation
    and Recreation
203 Governor Street, Suite 302
Richmond, VA 23219
804-786-7951

Virginia Division of State Parks
203 Governor Street, Suite 306
Richmond, VA 23219
804-786-1712

Virginia Wildlife Division
4010 W. Broad Street
Richmond, VA 23230
804-376-9588

Washington Area Bicyclist Association
1819 H Street, NW, Suite 640
Washington, DC 20006
202-872-9830

Washington Monument State Park
Route 1, Box 147
Middletown, MD 21769
301-432-8065

Watkins Regional Park
301 Watkins Park Drive
Upper Marlboro, MD 20772
301-249-9220

Westmoreland State Park
Route 1, Box 600
Montross, VA 22520
804-493-8821

Wheaton Regional Park
12012 Kemp Mill Road
Wheaton, MD 20902
301-946-7033

Wye Island Natural Resources
    Management Area
632 Wye Island Road
Queenstown, MD 21658
410-827-7577

Wye Oak State Park
13070 Crouse Mill Road
Queen Anne, MD 21657
410-634-2810

Appendix 3.
Common and Scientific Names of Plants and Animals

Common (English) names are used for plants and animals throughout this book. Few disciplines have accepted conventions for common names, and as a result, some common names refer to more than one species, and some species have been given more than one common name. Scientific (Latin) names are widely used in the scientific community and by serious amateur naturalists. This appendix gives the scientific name for plants and animals mentioned in this book. Note that advances in biology sometimes result in name changes for some species; such advances can be expected in the future.

Scientific names for individual species are composed of two words. The first is the genus and can be used independently. The second is a modifier that is used in tandem with the genus name to represent a given species. The notation *spp.* after a genus is used to represent several species of the genus. Some plants and animals mentioned in this book are identified only to the genus level. Other than butterflies and moths, insects are identified only to the coarse taxonomic level of order.

COMMON NAME	SCIENTIFIC NAME	COMMON NAME	SCIENTIFIC NAME
Woody plants		Azalea, Wild	*Rhododendron nudiflorum*
Ash, Green	*Fraxinus pennsylvanica*		
Ash, Pumpkin	*Fraxinus profunda*	Baldcypress	*Taxodium distichum*
Ash, Water	*Fraxinus caroliniana*	Basswood	*Tilia americana*
Ash, White	*Fraxinus americana*	Bay, Red	*Persea borbonia*
Aspen, Bigtooth	*Populus grandidentata*	Bay, Sweet	*Magnolia virginiana*
Azalea, Rosy Wild	*Rhododendron roseum*	Bayberry	*Myrica pennsylvanica*
		Beech, American	*Fagus grandifolia*

COMMON NAME	SCIENTIFIC NAME	COMMON NAME	SCIENTIFIC NAME
Birch, Black	*Betula lenta*	Magnolia, Southern	*Magnolia grandiflora*
Birch, River	*Betula nigra*	Magnolia, Swamp	*Magnolia virginiana*
Birch, Yellow	*Betula alleghaniensis*	Maple, Red	*Acer rubrum*
Blueberry	*Vaccinium* spp.	Maple, Silver	*Acer saccharinum*
Buttonbush	*Cephalanthus occidentalis*	Maple, Sugar	*Acer saccharum*
		Myrtle, Wax	*Myrica cerifera*
Cedar, Atlantic White	*Chamaecyparis thyoides*	Oak, Bear	*Quercus ilicifolia*
Cedar, Red	*Juniperus virginiana*	Oak, Blackjack	*Quercus marilandica*
Cherry, Black	*Prunus serotina*	Oak, Chestnut	*Quercus prinus*
Chestnut, American	*Castanea dentata*	Oak, Laurel	*Quercus laurifolia*
Chinquapin	*Castanea pumila*	Oak, Pin	*Quercus palustris*
Creeper, Trumpet	*Campsis radicans*	Oak, Post	*Quercus stellata*
Creeper, Virginia	*Parthenocissus quinquefolia*	Oak, Red	*Quercus rubra*
		Oak, Scarlet	*Quercus coccinea*
Crossvine	*Bignonia capreolata*	Oak, Southern Red	*Quercus falcata*
Dewberry	*Rubus permixtus*	Oak, Swamp Chestnut	*Quercus michauxii*
Dogwood, Flowering	*Cornus florida*	Oak, Water	*Quercus nigra*
Elder, Box	*Acer negundo*	Oak, White	*Quercus alba*
Elm, American	*Ulmus americana*	Oak, Willow	*Quercus phellos*
Elm, Slippery	*Ulmus rubra*	Pawpaw	*Asimina triloba*
Fetterbush	*Lyonia lucida*	Persimmon	*Diospyros virginiana*
Fir, Balsam	*Abies balsamea*	Pine, Loblolly	*Pinus taeda*
Fringe-tree	*Chionanthus virginica*	Pine, Pitch	*Pinus rigida*
Gallberry	*Ilex glabra*	Pine, Pond	*Pinus serotina*
Grape, Wild	*Vitis* spp.	Pine, Table Mountain	*Pinus pungens*
Greenbrier	*Smilax rotundifolia*	Pine, Virginia (Scrub)	*Pinus virginiana*
Gum, Black	*Nyssa sylvatica*	Pine, White	*Pinus strobus*
Gum, Swamp Black	*Nyssa biflora*	Pipevine	*Aristolochia macrophylla*
Gum, Tupelo	*Nyssa aquatica*		
Hackberry	*Celtis occidentalis*	Redbud	*Cercis canadensis*
Hazel, Witch	*Hamamelis virginiana*	Rhododendron	*Rhododendron maximum*
Heather, Beach	*Hudsonia tomentosa*		
Hemlock, Eastern	*Tsuga canadensis*	Rose, Multiflora	*Rosa multiflora*
Hickory, Mockernut	*Carya tomentosa*	Sassafras	*Sassafras albidum*
Hickory, Pignut	*Carya glabra*	Serviceberry	*Amelanchier laevis*
Hickory, Shagbark	*Carya ovata*	Spicebush	*Lindera benzoin*
Holly, American	*Ilex opaca*	Spruce, Red	*Picea rubens*
Honeysuckle, Japanese	*Lonicera japonica*	Sweetgum	*Liquidambar syraciflua*
Huckleberry	*Gaylussacia* spp.	Sycamore	*Platanus occidentalis*
Hydrangea, Climbing	*Decumaria barbara*	Tamarack	*Larix laricina*
Ivy, Poison	*Rhus radicans*	Teaberry	*Gaultheria procumbens*
Jessamine	*Gelsemium sempervirens*	Tree, Cucumber	*Magnolia acuminata*
		Tuliptree	*Liriodendron tulipifera*
Juniper, Virginia	*Juniperus virginiana*	Viburnum	*Viburnum* spp.
Laurel, Mountain	*Kalmia latifolia*	Walnut, Black	*Juglans nigra*
Laurel, Sheep	*Kalmia angustifolia*	Willow, Black	*Salix nigra*
Leucothoe	*Leucothoe axillaris*	Wintergreen, Common	*Gaultheria procumbens*
Locust, Black	*Robinia pseudo-acacia*		

COMMON NAME	SCIENTIFIC NAME	COMMON NAME	SCIENTIFIC NAME
Herbaceous plants		Cranberry	*Vaccinium macrocarpon*
		Cress, Rock	*Arabis laevigata*
Alexanders, Golden	*Zizia aurea*	Cress, Winter	*Barbarea vulgaris*
Alumroot	*Heuchera americana*	Dandelion	*Taraxacum officinale*
Anemone, Rue	*Anemonella thalictroides*	Dead-nettle, Purple	*Lamium purpureum*
		Dock, Water	*Rumex altissimus*
Anemone, Wood	*Anemone quinquefolia*	Duckweed	*Lemna* spp.
Arbutus, Trailing	*Epigaea repens*	Evening-primrose, Shale Barren	*Oenothera argillicola*
Arrowhead, Broad-leaved	*Sagittaria latifolia*	Fameflower	*Talinum teretifolium*
Arum, Arrow	*Peltandra virginica*	Fennel, Dog	*Eupatorium capillifolium*
Aster	*Aster* spp.		
Aster, Serpentine	*Aster depauperatus*	Flag, Sweet	*Acorus calamus*
Balm, Bee	*Monarda didyma*	Flower, Cardinal	*Lobelia cardinalis*
Beauty, Meadow	*Rhexia virginica*	Foamflower	*Tiarella cordifolia*
Beauty, Spring	*Claytonia virginica*	Gentian, Fringed	*Gentiana crinita*
Bellwort, Perfoliate	*Uvularia perfoliata*	Geranium, Wild	*Geranium maculatum*
Bindweed, Shale	*Convolvulus purshianus*	Ginger, Wild	*Asarum canadense*
		Ginseng, Dwarf	*Panax trifolius*
Bloodroot	*Sanguinaria canadensis*	Glasswort, Slender	*Salicornia europaea*
Bluebell, Virginia	*Mertensia virginica*	Goldenrod	*Solidago* spp.
Blueberry	*Vaccinium* spp.	Goldenrod, Seaside	*Solidago sempervirens*
Bluestem, Big	*Andropogon gerardii*	Goldthread	*Coptis groenlandica*
Bluestem, Little	*Andropogon scoparius*	Grass, Beach	*Ammophila breviligulata*
Breeches, Dutchman's	*Dicentra cucullaria*		
Bulrush, Great	*Scirpus validus*	Grass, Indian	*Sorghastrum nutans*
Bulrush, Saltmarsh	*Scirpus robustus*	Groundsel, Everlasting	*Senecio antennarifolius*
Bunchberry	*Cornus canadensis*	Harbinger-of-spring	*Erigenia bulbosa*
Burnet, Canada	*Sanguisorba canadensis*	Harperella	*Ptilimnium fluviatile*
Buttercup, Swamp	*Ranunculus hispidus*	Hawkweed	*Hieracium* spp.
Cabbage, Skunk	*Symplocarpus foetidus*	Hay, Saltmeadow	*Spartina patens*
Cattail, Broad-leaved	*Typha latifolia*	Heart, Bleeding	*Dicentra eximia*
Cattail, Narrow-leaved	*Typha angustifolia*	Henbit	*Lamium amplexicaule*
Chickweed, Common	*Stellaria media*	Hepatica	*Hepatica americana*
Chickweed, Star	*Stellaria pubera*	Hibiscus, Marsh	*Hibiscus moscheutos*
Cinquefoil, Rough-fruited	*Pontentilla norvegica*	Horsetail	*Equisetum* spp.
		Indigo, Blue False	*Baptisia australis*
Clintonia, White	*Clintonia umbellulata*	Ironweed, New York	*Vernonia noveboracensis*
Clover, Kates Mountain	*Trifolium virginicum*	Jack-in-the-pulpit	*Arisaema triphyllum*
		Lace, Queen Anne's	*Daucus carota*
Cohosh, Black	*Cimicifuga racemosa*	Ladyslipper, Pink	*Cypridedium acaule*
Cohosh, Blue	*Caulophyllum thalictroides*	Ladyslipper, Yellow	*Cypripedium calceolus*
		Lavender, Sea	*Limonium carolinianum*
Columbine	*Aquilegia canadensis*		
Coralroot, Spotted	*Corallorhiza maculata*	Leatherflower, Whitehaired	*Clematis albicoma*
Cordgrass, Big	*Spartina cynosuroides*		
Cordgrass, Saltmarsh	*Spartina alterniflora*	Leek, Wild (Ramp)	*Allium tricoccum*
Corn, Squirrel	*Dicentra canadensis*	Lettuce, Wild	*Lactuca canadensis*

COMMON NAME	SCIENTIFIC NAME	COMMON NAME	SCIENTIFIC NAME
Lily, Calla	*Calla palustris*	Saltgrass	*Distichlis spicata*
Lily, Canada	*Lilium canadense*	Salvia	*Salvia* spp.
Lizard's-tail	*Saururus cernuus*	Sandwort, Rock	*Arenaria stricta*
Mallow, Crimson-eyed Rose	*Hibiscus palustris*	Seal, False Solomon's	*Smilacina racemosa*
Mayapple	*Podophyllum peltatum*	Seal, Many-flowered Solomon's	*Polygonatum canaliculatum*
Mayflower, Canada	*Maianthemum canadense*	Seal, Solomon's	*Polygonatum biflorum*
Mermaids, False	*Floerkea proserpinacoides*	Sedge, Tussock	*Carex stricta*
		Sedge, Umbrella	*Cyperus esculentus*
Milkweed, Common	*Asclepias syriaca*	Shooting-star	*Dodecatheon meadia*
Milkweed, Green	*Asclepias viridiflora*	Skullcap, Showy	*Scutellaria serrata*
Milkweed, Red	*Asclepias rubra*	Spatterdock	*Nuphar advena*
Milkweed, Swamp	*Asclepias incarnata*	Speedwell	*Veronica* spp.
Millet, Walter's	*Echinochloa walteri*	Star, Blazing	*Liatris* spp.
Moss, Spanish	*Tillandsia usneoides*	Starflower	*Trientalis borealis*
Moss, Sphagnum	*Sphagnum* spp.	Stargrass, Yellow	*Hypoxis hirsuta*
Needlerush, Black	*Juncus roemerianus*	Stonecrop	*Sedum* spp.
Nettle, Stinging	*Urtica dioica*	Sundew, Round-leaved	*Drosera rotundifolia*
Onion, Shale	*Allium osyphilum*	Sunflower, Tickseed	*Bidens* spp.
Orchid, Cranefly	*Tipularia discolor*	Sunflower, Woodland	*Helianthus divaricatus*
Orchid, Round-leaved	*Habenaria orbiculata*	Susan, Black-eyed	*Rudbeckia hirta*
Orchid, Snake-mouth	*Pogonia ophioglossoides*	Switchgrass	*Panicum* spp.
Orchis, Showy	*Orchis spectabilis*	Tearthumb, Arrow-leaved	*Polygonum sagittatum*
Partridgeberry	*Mitchella repens*		
Petunia, Wild	*Ruellia* spp.	Thoroughwort	*Eupatorium altissimum*
Phacelia, Colville's	*Phacelia ranunculacea*	Three-square, American	*Scirpus pungens*
Phlox, Blue	*Phlox divaricata*		
Phlox, Moss	*Phlox subulata*	Three-square, Olney	*Scirpus americanus*
Phlox, Swordleaf	*Phlox buckleyi*	Toothwort, Cutleaf	*Dentaria laciniata*
Phlox, Wild Blue	*Phlox divaricata*	Trillium, Large-flowered	*Trillium grandiflorum*
Pickerelweed	*Pontederia cordata*		
Pimpernel, Mountain	*Pseudotaenidia montana*	Trillium, Painted	*Trillium undulatum*
		Trillium, Toadshade	*Trillium sessile*
Pink, Fire	*Silene virginica*	Trout-lily	*Erythronium americanum*
Pink, Grass	*Calopogon pulchellus*		
Plant, Pitcher	*Sarracenia purpurea*	Trout-lily, White	*Erythronium albidum*
Plantain, Rattlesnake	*Goodyera pubescens*	Turtlehead, White	*Chelone glabra*
Pogonia, Large Whorled	*Isotria verticillata*	Twayblade, Large	*Liparis lilifolia*
		Twayblade, Southern	*Listera australis*
Pogonia, Rose	*Pogonia ophioglossoides*	Twinleaf	*Jeffersonia diphylla*
Polygala, Fringed	*Polygala paucifolia*	Twisted-stalk, Rose	*Streptopus roseus*
Primrose, Evening	*Oenothera biennis*	Violet	*Viola* spp.
Puttyroot	*Aplectrum hyemale*	Violet, Birdsfoot	*Viola pedata*
Reed, Common	*Phragmites australis*	Violet, Green	*Hybanthus concolor*
Rocket, Sea	*Cakile edentula*	Waterleaf, Virginia	*Hydrophyllum virginianum*
Rush, Needle	*Juncus* spp.		
Rush, Soft	*Juncus effusus*	Weed, Butterfly	*Asclepias tuberosa*

COMMON NAME	SCIENTIFIC NAME	COMMON NAME	SCIENTIFIC NAME
Weed, Joe-pye	*Eupatorium purpureum*	Meadow Mushroom	*Agaricus campestris*
Weed, Rattlesnake	*Hieracium venosum*	Morel	*Morchella* spp.
Wildcelery	*Vallisneria spiralis*	Orange Peel	*Aleuria aurantia*
Wildrice	*Zizania aquatica*	Russula	*Russula* spp.
Wintergreen	*Gaultheria procumbens*	Scarlet Cup	*Sarcoscypha coccinea*
Wintergreen, Spotted	*Chimaphila maculatum*	Sulphur Shelf	*Laetiporus sulphureus*
		Turkey-tails	*Coriolus versicolor*
		Witches' Butter	*Tremella mesenterica*

Ferns

Cliffbrake, Purple	*Pellaea atropurpurea*		
Fern, Allegheny Cliff	*Woodsia scopulina*	**Mammals**	
Fern, Bracken	*Pteridium aquilinum*	Bat, Big Brown	*Eptesicus fuscus*
Fern, Broad Beech	*Phegopteris hexagonoptera*	Bat, Evening	*Nycticeius humeralis*
		Bat, Indiana	*Myotis sodalis*
Fern, Christmas	*Polystichum acrostichoides*	Bat, Red	*Lasiurus borealis*
Fern, Cinnamon	*Osmunda cinnamomea*	Bat, Townsend's Big-eared	*Plecotus townsendii*
Fern Curly-grass	*Schizaea pusilla*	Bear, Black	*Ursus americanus*
Fern, Hairy Lip	*Cheilanthes lanosa*	Beaver	*Castor canadensis*
Fern, Hay-scented	*Dennstaedtia punctilobula*	Bison	*Bison bison*
		Bobcat	*Lynx rufus*
Fern, Intermediate Wood	*Dryopteris intermedia*	Chipmunk, Eastern	*Tamias striatus*
		Coyote	*Canis latrans*
Fern, Interrupted	*Osmunda claytoniana*	Deer, White-tailed	*Odocoileus virginianus*
Fern, Lady	*Athyrium filix-femina*	Dolphin, Atlantic Bottle-nosed	*Tursiops truncatus*
Fern, Marginal Shield	*Dryopteris marginalis*		
Fern, New York	*Thelypteris noveboracensis*	Elk	*Cervus elaphus*
		Fox, Gray	*Urocyon cinereoargenteus*
Fern, Royal	*Osmunda regalis*		
Fern, Sensitive	*Onoclea sensibilis*	Fox, Red	*Vulpes vulpes*
Fern, Walking	*Camptosorus rhizophyllus*	Lion, Mountain	*Felis concolor*
		Mink	*Mustela vison*
Polypody, Rock	*Polypodium virginianum*	Mouse, Deer	*Peromyscus maniculatus*
		Mouse, Harvest	*Reithrodontomys humulis*
Spleenwort, Ebony	*Asplenium platyneuron*		
Spleenwort, Mountain	*Asplenium montanum*	Mouse, House	*Mus musculus*
Wallrue, American	*Asplenium ruta-muraria*	Mouse, White-footed	*Peromyscus leucopus*
		Muskrat	*Ondatra zibethicus*
		Myotis, Little Brown	*Myotis lucifugus*
		Nutria	*Myocastor coypus*
Fungi		Opossum	*Didelphis virginiana*
		Otter, River	*Lutra canadensis*
Cedar Apple Rust	*Gymnosporangium juniperi-virginiana*	Pipistrelle, Eastern	*Pipistrellus subflavus*
		Rabbit, Eastern Cottontail	*Sylvilagus floridanus*
Dead Man's Fingers	*Xylaria polymorpha*		
Earthstar	*Geastrum* spp.	Rabbit, Marsh	*Sylvilagus palustris*
Golden Chanterelle	*Cantharellus lutescens*	Raccoon	*Procyon lotor*
Inky Cap	*Coprinus* spp.	Rat, Black	*Rattus rattus*
Jack-o-lantern	*Omphalotus illudens*		

COMMON NAME	SCIENTIFIC NAME	COMMON NAME	SCIENTIFIC NAME
Rat, Marsh Rice	*Oryzomys palustris*	Snake, Black Rat	*Elaphe obsoleta*
Rat, Norway	*Rattus norvegicus*	Snake, Brown	*Storeria dekayi*
Seal, Harbor	*Phoca vitulina*	Snake, Corn	*Elaphe guttata*
Seal, Harp	*Phoca groenlandica*	Snake, Eastern Garter	*Thamnophis sirtalis*
Seal, Hooded	*Cystophora cristata*	Snake, Hognose	*Heterodon platyrhinos*
Skunk, Spotted	*Spilogale putorius*	Snake, Northern Water	*Natrix sipedon*
Skunk, Striped	*Mephitis mephitis*		
Squirrel, Eastern Gray	*Sciurus carolinensis*	Snake, Redbelly	*Storeria occipitomaculata*
Squirrel, Fox	*Sciurus niger*		
Squirrel, Red	*Tamiasciurus hudsonicus*	Snake, Ring-necked	*Diadophis punctatus*
		Snake, Rough Green	*Opheodrys aestivus*
Squirrel, Southern Flying	*Glaucomys volans*	Snake, Scarlet	*Cemophora coccinea*
		Snake, Smooth Earth	*Virginia valeriae*
Vole, Meadow	*Microtus pennsylvanicus*	Snake, Worm	*Carphophis amoenus*
		Stinkpot	*Sternotherus odoratus*
Vole, Rock	*Microtus chrotorrhinus*	Terrapin, Diamondback	*Malaclemys terrapin*
Vole, Southern Red-backed	*Clethrionomys gapperi*		
		Turtle, Box	*Terrapene carolina*
Vole, Woodland	*Microtus pinetorum*	Turtle, Eastern Mud	*Kinosternon subrubrum*
Weasel, Least	*Mustela nivalis*	Turtle, Loggerhead Sea	*Caretta caretta*
Weasel, Long-tailed	*Mustela frenata*	Turtle, Painted	*Chrysemys picta*
Woodchuck (Groundhog)	*Marmota monax*	Turtle, Snapping	*Chelydra serpentina*
		Turtle, Spotted	*Clemmys guttata*
Woodrat, Eastern	*Neotoma floridana*	Turtle, Yellow-bellied	*Chrysemys scripta*

Reptiles

Amphibians

COMMON NAME	SCIENTIFIC NAME	COMMON NAME	SCIENTIFIC NAME
Anole, Green	*Anolis carolinensis*	Bullfrog	*Rana catesbeiana*
Cooter, River	*Chrysemys concinna*	Frog, Carpenter	*Rana virgatipes*
Copperhead	*Agkistrodon contortrix*	Frog, Green	*Rana clamitans*
Cottonmouth	*Agkistrodon piscivorus*	Frog, Pickerel	*Rana palustris*
Kingsnake, Eastern	*Lampropeltis getulus*	Frog, Wood	*Rana sylvatica*
Kingsnake, Scarlet	*Lampropeltis triangulum*	Newt, Red-spotted (Red Eft)	*Notophthalmus viridescens*
Lizard, Eastern Fence	*Sceloporus undulatus*	Peeper, Spring	*Hyla crucifer*
Lizard, Slender Glass	*Ophisaurus attenuatus*	Salamander, Appalachian Seal	*Desmognathus monticola*
Racerunner, Six-lined	*Cnemidophorus sexlineatus*	Salamander, Cheat Mountain	*Plethodon nettingi*
Rattlesnake, Canebrake	*Crotalus adamanteus*	Salamander, Green	*Aneides aeneus*
Rattlesnake, Timber	*Crotalus horridus*	Salamander, Jefferson	*Ambystoma jeffersonianum*
Skink, Broad-headed	*Eumeces laticeps*		
Skink, Coal	*Eumeces anthracinus*	Salamander, Long-tailed	*Eurycea longicauda*
Skink, Five-lined	*Eumeces fasciatus*		
Skink, Ground	*Scincella lateralis*	Salamander, Marbled	*Ambystoma opacum*
Skink, Southeastern Five-lined	*Eumeces inexpectatus*	Salamander, Mud	*Pseudotriton montanus*
		Salamander, Northern Dusky	*Desmognathus fuscus*
Slider, Red-bellied	*Chrysemys rubriventris*		

COMMON NAME	SCIENTIFIC NAME	COMMON NAME	SCIENTIFIC NAME
Salamander, Peaks of Otter	*Plethodon hubrichti*	Cormorant, Great	*Phalacrocorax carbo*
Salamander, Red	*Pseudotriton ruber*	Crane, Whooping	*Grus americana*
Salamander, Red-backed	*Plethodon cinereus*	Creeper, Brown	*Certhia americana*
		Crossbill, Red	*Loxia curvirostra*
Salamander, Shenandoah	*Plethodon shenandoah*	Crossbill, White-winged	*Loxia leucoptera*
Salamander, Slimy	*Plethodon glutinosus*	Crow, American	*Corvus brachyrhynchos*
Salamander, Spotted	*Ambystoma maculatum*	Crow, Fish	*Corvus ossifragus*
Salamander, Spring	*Gyrinophilus porphyriticus*	Cuckoo, Black-billed	*Coccyzus erythropthalmus*
Salamander, Three-lined	*Eurycea guttonlineata*	Cuckoo, Yellow-billed	*Coccyzus americanus*
		Dove, Mourning	*Zenaida macroura*
Salamander, Tiger	*Ambystoma tigrinum*	Duck, Harlequin	*Histrionicus histrionicus*
Salamander, Two-lined	*Eurycea bislineata*	Duck, Ring-necked	*Aythya collaris*
		Duck, Ruddy	*Osyura jamaicensis*
Salamander, Wehrle's	*Plethodon wehrlei*	Duck, Wood	*Aix sponsa*
Toad, American	*Bufo americana*	Dunlin	*Calidris alpina*
Toad, Fowler's	*Bufo woodhousei*	Eagle, Bald	*Haliaeetus leucocephalus*
Treefrog, Gray	*Hyla versicolor*		
Treefrog, Green	*Hyla cinerea*	Eagle, Golden	*Aquila chrysaetos*
Treefrog, Pine Barrens	*Hyla andersoni*	Egret, Great	*Casmerodius albus*
		Egret, Reddish	*Egretta rufescens*
		Egret, Snowy	*Egretta thula*
Birds		Falcon, Peregrine	*Falco peregrinus*
		Finch, House	*Carpodacus mexicanus*
Bittern, American	*Botaurus lentiginosus*	Finch, Purple	*Carpodacus purpureus*
Bittern, Least	*Ixobrychus exilis*	Flicker, Northern	*Colaptes auratus*
Blackbird, Red-winged	*Agelaius phoeniceus*	Flycatcher, Acadian	*Empidonax virescens*
		Flycatcher, Alder	*Empidonax alnorum*
Blackbird, Rusty	*Euphagus carolinus*	Flycatcher, Fork-tailed	*Tyrannus savana*
Bluebird, Eastern	*Sialia sialis*	Flycatcher, Least	*Empidonax minimus*
Bobolink	*Dolichonyx oryzivorus*	Flycatcher, Willow	*Empidonax traillii*
Bobwhite, Northern	*Colinus virginianus*	Flycatcher, Yellow-bellied	*Empidonax flaviventris*
Brant	*Branta bernicla*		
Bufflehead	*Bucephala albeola*	Gannet, Northern	*Morus bassanus*
Bunting, Indigo	*Passerina cyanea*	Goldeneye, Common	*Bucephala clangula*
Bunting, Snow	*Plectrophenax nivalis*	Goldfinch, American	*Carduelis tristis*
Canvasback	*Aythya valisineria*	Goose, Canada	*Branta canadensis*
Cardinal, Northern	*Cardinalis cardinalis*	Goose, Ross'	*Chen rossii*
Catbird, Gray	*Dumetella carolinensis*	Goose, Snow	*Chen caerulescens*
Chat, Yellow-bellied	*Icteria virens*	Goshawk, Northern	*Accipiter gentilis*
Chickadee, Black-capped	*Parus atricapillus*	Grebe, Eared	*Podiceps nigricollis*
		Grebe, Horned	*Podiceps auritus*
Chickadee, Carolina	*Parus carolinensis*	Grebe, Pied-billed	*Podilymbus podiceps*
Chuck-will's-widow	*Caprimulgus carolinensis*	Grebe, Red-necked	*Podiceps grisegena*
		Grosbeak, Blue	*Guiraca caerulea*
Coot, American	*Fulica americana*	Grosbeak, Evening	*Coccothraustes vespertinus*
Cormorant, Double-crested	*Phalacrocorax auritus*		

COMMON NAME	SCIENTIFIC NAME	COMMON NAME	SCIENTIFIC NAME
Grosbeak, Rose-breasted	*Pheucticus ludovicianus*	Merlin	*Falco columbarius*
Grouse, Ruffed	*Bonasa umbellus*	Mockingbird, Northern	*Mimus polyglottos*
Gull, Black-headed	*Larus ridibundus*	Nighthawk, Common	*Chordeiles minor*
Gull, Bonaparte's	*Larus philadelphia*	Night-Heron, Black-crowned	*Nycticorax nycticorax*
Gull, Glaucous	*Larus hyperboreus*		
Gull, Great Black-backed	*Larus marinus*	Night-Heron, Yellow-crowned	*Nyctanassa violacea*
Gull, Herring	*Larus argentatus*	Nuthatch, Brown-headed	*Sitta pusilla*
Gull, Iceland	*Larus glaucoides*		
Gull, Laughing	*Larus atricilla*	Nuthatch, Red-breasted	*Sitta canadensis*
Gull, Lesser Black-backed	*Larus fuscus*		
Gull, Little	*Larus minutus*	Nuthatch, White-breasted	*Sitta carolinensis*
Gull, Ring-billed	*Larus delawarensis*	Oldsquaw	*Clangula hyemalis*
Gull, Ross'	*Rhodostethia rosea*	Oriole, Baltimore	*Icterus galbula*
Gull, Thayer's	*Larus thayeri*	Oriole, Orchard	*Icterus spurius*
Gull, Yellow-legged	*Larus cachinnans*	Osprey	*Pandion haliaetus*
Harrier, Northern	*Circus cyaneus*	Ovenbird	*Seiurus aurocapillus*
Hawk, Broad-winged	*Buteo platypterus*	Owl, Barn	*Tyto alba*
Hawk, Cooper's	*Accipiter cooperii*	Owl, Barred	*Strix varia*
Hawk, Red-shouldered	*Buteo lineatus*	Owl, Great Horned	*Bubo virginianus*
		Owl, Northern Saw-whet	*Aegolius acadicus*
Hawk, Red-tailed	*Buteo jamaicensis*		
Hawk, Rough-legged	*Buteo lagopus*	Owl, Short-eared	*Asio flammeus*
Hawk, Sharp-shinned	*Accipiter striatus*	Owl, Snowy	*Nyctea scandica*
Heron, Great Blue	*Ardea herodias*	Oystercatcher, American	*Haematopus palliatus*
Heron, Green	*Butorides striatus*		
Heron, Little Blue	*Egretta caerulea*	Parakeet, Carolina	*Conuropsis carolinensis*
Hummingbird, Ruby-throated	*Archilochus colubris*	Parula, Northern	*Parula americana*
		Pelican, Brown	*Pelecanus occidentalis*
Ibis, Glossy	*Plegadis falcinellus*	Pheasant, Ring-necked	*Phasianus colchicus*
Ibis, White	*Eudocimus albus*		
Junco, Dark-eyed	*Junco hyemalis*	Phoebe, Eastern	*Sayornis phoebe*
Kestrel, American	*Falco sparverius*	Pigeon, Passenger	*Ectopistes migratorius*
Kingfisher, Belted	*Ceryle alcyon*	Pintail, Northern	*Anas acuta*
Kinglet, Golden-crowned	*Regulus satrapa*	Pipit, American	*Anthus rubescens*
		Plover, Piping	*Charadrius melodus*
Kinglet, Ruby-crowned	*Regulus calendula*	Rail, Black	*Laterallus jamaicensis*
		Rail, Clapper	*Rallus longirostris*
Knot, Red	*Calidris canutus*	Rail, King	*Rallus elegans*
Limpkin	*Aramus guarauna*	Rail, Virginia	*Rallus limicola*
Loon, Common	*Gavia immer*	Raven, Common	*Corvus corax*
Loon, Red-throated	*Gavia stellata*	Redpoll, Common	*Carduelis flammea*
Mallard	*Anas platyrhynchos*	Redpoll, Hoary	*Carduelis hornemanni*
Meadowlark, Eastern	*Sturnella magna*	Redstart, American	*Setophaga ruticilla*
Merganser, Common	*Mergus merganser*	Robin, American	*Turdus migratorius*
Merganser, Red-breasted	*Mergus serrator*	Ruff	*Philomachus pugnax*
		Sanderling	*Calidris alba*

COMMON NAME	SCIENTIFIC NAME	COMMON NAME	SCIENTIFIC NAME
Sandpiper, Curlew	*Calidris ferruginea*	Swift, Chimney	*Chaetura pelagica*
Sandpiper, Purple	*Calidris maritima*	Tanager, Scarlet	*Piranga olivacea*
Sapsucker, Yellow-bellied	*Sphyrapicus varius*	Tanager, Summer	*Piranga rubra*
		Teal, Blue-winged	*Anas discors*
Scaup, Greater	*Aythya marila*	Teal, Green-winged	*Anas crecca*
Scaup, Lesser	*Aythya affinis*	Tern, Black	*Chlidonias niger*
Scoter, Black	*Melanitta nigra*	Tern, Common	*Sterna hirundo*
Scoter, Surf	*Melanitta perspicillata*	Tern, Least	*Sterna antillarum*
Scoter, White-winged	*Melanitta fusca*	Tern, Royal	*Sterna maxima*
		Tern, Whiskered	*Chlidonias hybridus*
Screech-Owl, Eastern	*Otus asio*	Tern, White-winged	*Chlidonias leucopterus*
Shoveler, Northern	*Anas clypeata*	Thrush, Gray-cheeked	*Catharus minimus*
Shrike, Loggerhead	*Lanius ludovicianus*	Thrush, Hermit	*Catharus guttatus*
Shrike, Northern	*Lanius excubitor*	Thrush, Swainson's	*Catharus ustulatus*
Siskin, Pine	*Carduelis pinus*	Thrush, Wood	*Hylocichla mustelina*
Skimmer, Black	*Rynchops niger*	Titmouse, Tufted	*Parus bicolor*
Snipe, Common	*Gallinago gallinago*	Towhee, Eastern	*Pipilo erythrophthalmus*
Sora	*Porzana carolina*		
Sparrow, American Tree	*Spizella arborea*	Turkey, Wild	*Meleagris gallopavo*
		Veery	*Catharus fuscescens*
Sparrow, Field	*Spizella pusilla*	Vireo, Red-eyed	*Vireo olivaceus*
Sparrow, Fox	*Passerella iliaca*	Vireo, Solitary	*Vireo solitarius*
Sparrow, Grasshopper	*Ammodramus savannarum*	Vireo, White-eyed	*Vireo griseus*
		Vireo, Yellow-throated	*Vireo flavifrons*
Sparrow, Henslow's	*Ammodramus henslowii*		
Sparrow, House	*Passer domesticus*	Warbler, Black-and-white	*Mniotilta varia*
Sparrow, Ipswich	*Passerculus sandwichensis princeps*	Warbler, Black-throated Blue	*Dendroica caerulescens*
Sparrow, Saltmarsh Sharp-tailed	*Ammodramus caudacutus*	Warbler, Black-throated Green	*Dendroica virens*
Sparrow, Savannah	*Passerculus sandwichensis*	Warbler, Blackburnian	*Dendroica fusca*
Sparrow, Seaside	*Ammodramus maritimus*	Warbler, Canada	*Wilsonia canadensis*
		Warbler, Cape May	*Dendroica tigrina*
Sparrow, Song	*Melospiza melodia*	Warbler, Cerulean	*Dendroica cerulea*
Sparrow, Swamp	*Melospiza georgiana*	Warbler, Chestnut-sided	*Dendroica pensylvanica*
Sparrow, White-throated	*Zonotrichia albicollis*	Warbler, Golden-winged	*Vermivora chrysoptera*
Starling, European	*Sturnus vulgaris*	Warbler, Hooded	*Wilsonia citrina*
Stint, Red-necked	*Calidris ruficollis*	Warbler, Kentucky	*Oporornis formosus*
Swallow, Barn	*Hirundo rustica*	Warbler, Magnolia	*Dendroica magnolia*
Swallow, Northern Rough-winged	*Stelgidopteryx serripennis*	Warbler, Mourning	*Oporornis philadelphia*
		Warbler, Nashville	*Vermivora ruficapilla*
Swallow, Tree	*Tachycineta bicolor*	Warbler, Orange-crowned	*Vermivora celata*
Swan, Trumpeter	*Cygnus buccinator*		
Swan, Tundra	*Cygnus columbianus*	Warbler, Palm	*Dendroica palmarum*

COMMON NAME	SCIENTIFIC NAME	COMMON NAME	SCIENTIFIC NAME
Warbler, Pine	*Dendroica pinus*	Chub, Creek	*Semotilus atromaculatus*
Warbler, Prairie	*Dendroica discolor*		
Warbler, Prothonotary	*Protonotaria citrea*	Croaker, Atlantic	*Micropogonias undulatus*
Warbler, Swainson's	*Limnothlypis swainsonii*	Dace, Black-nosed	*Rhinichthys atratulus*
Warbler, Tennessee	*Vermivora peregrina*	Dace, Rosy-sided	*Clinostomus funduloides*
Warbler, Worm-eating	*Helmitheros vermivorus*		
Warbler, Yellow	*Dendroica petechia*	Darter, Fantail	*Etheostoma flabellare*
Warbler, Yellow-rumped (Myrtle)	*Dendroica coronata*	Darter, Shield	*Percina peltata*
		Darter, Tessellated	*Etheostoma olmstedi*
Warbler, Yellow-throated	*Dendroica dominica*	Drum, Black	*Pogonias cromis*
		Drum, Red	*Sciaenops ocellatus*
Waterthrush, Louisiana	*Seiurus motacilla*	Flounder	*Paralichthys dentatus*
		Killifish, Banded	*Fundulus diaphanus*
Waterthrush, Northern	*Seiurus noveboracensis*	Killifish, Rainwater	*Lucania parva*
		Menhaden, Atlantic	*Brevoortia tyrannus*
Waxwing, Cedar	*Bombycilla cedrorum*	Minnow, Fathead	*Pimephales promelas*
Whimbrel	*Numenius phaeopus*	Minnow, Sheepshead	*Cyprinodon variegatus*
Whip-poor-will	*Caprimulgus vociferus*	Mummichog	*Fundulus heteroclitus*
Wigeon, American	*Anas americana*	Perch, White	*Morone americana*
Wigeon, Eurasian	*Anas penelope*	Perch, Yellow	*Perca flavescens*
Willet	*Catoptrophorus semipalmatus*	Sculpin, Mottled	*Cottus bairdi*
		Seatrout, Spotted	*Cynoscion nebulosus*
Woodcock, American	*Scolopax minor*	Shad, Atlantic	*Alosa sapidissima*
Woodpecker, Downy	*Picoides pubescens*	Spot	*Leiostomus xanthurus*
Woodpecker, Hairy	*Picoides villosus*	Stickleback, Fourspine	*Apeltes quadracus*
Woodpecker, Pileated	*Dryocopus pileatus*	Sunfish, Pumpkinseed	*Lepomis gibbosus*
Woodpecker, Red-bellied	*Melanerpes carolinus*	Trout, Brook	*Salvelinus fontinalis*
		Trout, Brown	*Salmo trutta*
Woodpecker, Red-headed	*Melanerpes erythrocephalus*	Trout, Rainbow	*Oncorhynchus mykiss*
Wood-Pewee, Eastern	*Contopus virens*		
Wren, Carolina	*Thryothorus ludovicianus*		

Insects (identified only to order)

COMMON NAME	SCIENTIFIC NAME
Wren, House	*Troglodytes aedon*
Wren, Marsh	*Cistothorus palustris*
Wren, Winter	*Troglodytes troglodytes*
Yellowthroat, Common	*Geothlypis trichas*

Bees, Wasps, and Ants	Order Hymenoptera
Beetles	Order Coleoptera
Butterflies and Moths	Order Lepidoptera
Caddisflies	Order Trichoptera
Cicadas, Leafhoppers, and Aphids	Order Homoptera
Crickets, Grass-hoppers, Mantids, Walking Sticks, Katydids, Roaches	Order Orthoptera

Fish

COMMON NAME	SCIENTIFIC NAME
Bass, Black Sea	*Centropristis striata*
Bass, Largemouth	*Micropterus salmoides*
Bass, Smallmouth	*Micropterus dolomieu*
Bass, Striped	*Morone saxatilis*
Bluefish	*Pomatomus saltatrix*

Damselflies and Dragonflies	Order Odonata
Flies	Order Diptera
Lacewings, Antlions, Fishflies	Order Neuroptera

COMMON NAME	SCIENTIFIC NAME	COMMON NAME	SCIENTIFIC NAME
Mayflies	Order Ephemeroptera	Marble, Olympia	*Euchloe olympia*
Stoneflies	Order Plecoptera	Mark, Question	*Polygonia interrogationis*
True Bugs	Order Hemiptera	Metalmark, Northern	*Calephelis borealis*
		Monarch	*Danaus plexippus*
		Moth, Luna	*Actias luna*
Butterflies and Moths		Moth, Sphinx	Family Sphingidae
Bear, Wooly	*Isia isabella*	Nymph, Common Wood	*Cercyonis pegala*
Blue, Dusky	*Celastrina ebenina*		
Blue, Eastern Tailed	*Everes comyntas*	Orange-tip, Falcate	*Anthocharis midea*
Blue, Silvery	*Glaucopsyche lygdamus*	Purple, Red-spotted	*Limenitis arthemis astyanax*
Buckeye	*Junonia coenia*		
Checkerspot, Baltimore	*Euphydryas phaeton*	Sachem	*Atalopedes campestris*
		Satyr, Little Wood	*Megisto cymela*
Checkerspot, Harris'	*Chlosyne harrisii*	Skipper, Clouded	*Lerema accius*
Checkerspot, Silvery	*Chlosyne nycteis*	Skipper, Common Checkered	*Pyrgus communis*
Cloak, Mourning	*Nymphalis antiopa*		
Comma, Eastern	*Polygonia comma*	Skipper, Fiery	*Hylephila phyleus*
Copper, American	*Lycaena phlaeas*	Skipper, Gold-banded	*Autochton cellus*
Crescent, Pearl	*Phyciodes tharos*	Skipper, Hobomok	*Poanes hobomok*
Devil, Hickory Horn	*Citheronia regalis*	Skipper, Leonard's	*Hesperia leonardus*
Duskywing, Juvenal's	*Erynnis juvenalis*	Skipper, Long-tailed	*Urbanus proteus*
Elfin, Frosted	*Incisalia irus*	Skipper, Ocola	*Panoquina ocola*
Elfin, Henry's	*Callophrys henrici*	Skipper, Rare	*Problema bulenta*
Elfin, Pine	*Incisalia niphon*	Skipper, Salt Marsh	*Panoquina panoquin*
Emperor, Tawny	*Asterocampa clyton*	Skipper, Silver Spotted	*Epargyreus clarus*
Eye, Northern Pearly	*Enodia anthedon*		
Fritillary, Aphrodite	*Speyeria aphrodite*	Snout, American	*Libytheana carinenta*
Fritillary, Atlantis	*Speyeria atlantis*	Sulphur, Clouded	*Colias philodice*
Fritillary, Diana	*Speyeria diana*	Sulphur, Cloudless	*Phoebis sennae*
Fritillary, Great Spangled	*Speyeria cybele*	Sulphur, Pink-edged	*Colias interior*
		Swallowtail, Eastern Tiger	*Papilio glaucus*
Fritillary, Meadow	*Boloria bellona*		
Fritillary, Silver-bordered	*Boloria selene*	Swallowtail, Palamedes	*Papilio palamedes*
Hackberry	*Asterocampa celtis*	Swallowtail, Pipevine	*Battus philenor*
Hairstreak, Coral	*Satyrium titus*	Swallowtail, Spicebush	*Papilio troilus*
Hairstreak, Great Purple	*Atlides halesus*	Swallowtail, Zebra	*Eurytides marcellus*
Hairstreak, Olive	*Mitoura grynea*	White, Cabbage	*Pieris rapae*
Hairstreak, White-M	*Parrhasium m-album*	White, West Virginia	*Pieris virginiensis*
Harvester	*Feniseca tarquinius*		

References

Adkins, Leonard M. 1991. *Walking the Blue Ridge.* Chapel Hill: University of North Carolina Press.

Anderson, Bill. 1991. *Fishing the rivers of the mid-Atlantic.* Centreville, Md.: Tidewater Publishers.

Arora, David. 1986. *Mushrooms demystified.* Berkeley: Ten Speed Press.

Badger, Curtis J. 1994. *A birdwatcher's guide to Virginia's Eastern Shore.* Onley, Va.: Salicornia Press.

Baliles, Gerald L. 1995. *Preserving the Chesapeake Bay.* McLean, Va.: EPM Publications.

Baron, Robert C., and Elizabeth Darby Junkin, eds. 1986. *Of discovery and destiny.* Golden, Colo.: Fulcrum Publishing.

Bartram, William. 1980. *Travels.* Salt Lake City: Peregrine Smith.

Beatty, Christopher, 1995. *The Susquehanna River guide.* Landenberg, Pa.: Ecopress.

Berman, Richard L., and Deborah Gerhard. 1994. *Natural Washington.* McLean, Va.: EPM Publications.

Bishop, Sherman C. 1943. *Handbook of salamanders.* Ithaca, N.Y.: Comstock Publishing.

Bonta, Marcia. 1991. *Appalachian spring.* Pittsburgh: University of Pittsburgh Press.

————. 1994. *Appalachian autumn.* Pittsburgh: University of Pittsburgh Press.

Borror, Donald J., and Richard E. White. 1970. *A field guide to the insects of America north of Mexico.* Boston: Houghton Mifflin.

Bowen, John. 1990. *Adventuring in the Chesapeake Bay area.* San Francisco: Sierra.

Braun, E. Lucy. 1950. *Deciduous forests of North America.* Philadelphia: Blackiston.

Briggs, Shirley A., ed. 1954. *Washington: City in the woods.* Washington, D.C.: Audubon Society of the District of Columbia.

Britton, Nathaniel Lord, and Addison Brown. 1970. *An illustrated flora of the northern United States and Canada.* 3 vols. New York: Dover.

249

Brockman, C. Frank, Rebecca Merrilees, and Herbert S. Zim. 1968. *Trees of North America*. New York: Golden Press.

Brooks, Maurice. 1965. *The Appalachians*. Morgantown, W.Va.: Seneca Books.

Brooks, William K. 1996. *The oyster*. Baltimore: Johns Hopkins University Press.

Brown, Lauren. 1979. *Grasses, an identification guide*. Boston: Houghton Mifflin.

Brown, Melvin L., and Russell G. Brown. 1984. *Herbaceous plants of Maryland*. Baltimore: Port City Press.

Brown, Russell G., and Melvin L. Brown. 1972. *Woody plants of Maryland*. Baltimore: Port City Press.

Brown, Tom, Jr., and Brandt Morgan. 1983. *Tom Brown's field guide to nature observation and tracking*. New York: Berkley Books.

Buckelew, Albert R., Jr., and George A. Hall. 1994. *The West Virginia breeding bird atlas*. Pittsburgh: University of Pittsburgh Press.

Bulloch, David K. 1991. *The American Littoral Society handbook for the marine naturalist*. New York: Walker & Co.

Burger, Joanna. 1996. A naturalist along the Jersey shore. New Brunswick, N.J.: Rutgers University Press.

Burt, William Henry. 1976. *A field guide to the mammals*. Boston: Houghton Mifflin.

Carr, Martha S. 1950. *The District of Columbia: Its rocks and their geologic history*. U.S. Geological Survey Bulletin 967. Washington, D.C.: U.S. Geological Survey.

Carson, Rachel L. 1950. *The sea around us*. New York: Oxford University Press.

———. 1962. *Silent spring*. Boston: Houghton Mifflin.

Catlin, D. T. 1984. *A naturalist's Blue Ridge Parkway*. Knoxville: University of Tennessee Press.

Choukas-Bradley, Melanie, and Polly Alexander. 1987. *City of trees*. Baltimore: Johns Hopkins University Press.

Clarkson, Roy B. 1964. *Tumult on the mountains*. Parsons, W. Va.: McClain.

Cobb, Boughton. 1963. *A field guide to the ferns and their related families*. Boston: Houghton Mifflin.

Collins, Henry Hill, Jr. 1959. *Complete field guide to American wildlife*. New York: Harper and Brothers.

Conant, Roger, and Joseph T. Collins. 1991. *A field guide to reptiles and amphibians: Eastern and central North America*. Boston: Houghton Mifflin.

Connor, Jack. 1988. *The complete birder*. Boston: Houghton Mifflin Company.

———. 1991. *Season at the point*. New York: Atlantic Monthly Press.

Conners, John A. 1988. *Shenandoah National Park: An interpretive guide*. Blacksburg, Va.: McDonald & Woodward Publishing.

Core, Earl L. 1966. *Vegetation of West Virginia*. Parsons, W. Va.: McClain.

———. 1981. *Spring wild flowers of West Virginia*. Morgantown: West Virginia University Press.

Core, Earl L., and Nelle P. Ammons. 1958. *Woody plants in winter*. Pacific Grove, Ca.: Boxwood.

Covell, Charles V., Jr. 1984. *A field guide to the moths of eastern North America.* Boston: Houghton Mifflin.

Davis, Hubert J. 1962. *The Great Dismal Swamp.* Richmond, Va.: Cavalier Press.

de Hart, Allen, and Bruce Sundquist. 1993. *Monongahela National Forest hiking guide.* Charleston: West Virginia Highlands Conservancy.

Dietrich, R. V. 1970. *Geology and Virginia.* Charlottesville: University Press of Virginia.

Ehrlich, Paul R., David S. Dobkin, and Darryl Wheye. 1988. *The birder's handbook.* New York: Fireside Books.

Eleuterius, Lionel N. 1990. *Tidal marsh plants.* Gretna, La.: Pelican.

Elman, Robert. 1977. *First in the field: America's pioneering naturalists.* New York: Van Nostrand Reinhold.

Evans, Howard Ensign. 1993. *Pioneer naturalists.* New York: Henry Holt & Company.

Fernald, Merritt Lyndon. 1970. *Gray's manual of botany.* New York: Van Nostrand Reinhold.

Fisher, Alan. 1984. *Country walks near Washington.* Boston: Appalachian Mountain Club.

Fisher, Alan. 1985. *More country walks near Washington.* Baltimore: Rambler Books.

Fisher, George W., F. J. Pettijohn, J. C. Reed, Jr., and Kenneth Weaver, eds. 1970. *Studies of Appalachian geology: Central and southern.* New York: Wiley.

Fleming, Cristol, Marion Blos Lobstein, and Barbara Tufty. 1995. *Finding wildflowers in the Washington-Baltimore area.* Baltimore: Johns Hopkins University Press.

Flynn, Kevin C., and Elizabeth S. Wooster. 1995. *The Potomac River guide.* Kensington, Md.: Flynn-Wooster Editorial Services.

Frye, Keigh. 1986. *Roadside geology of Virginia.* Missoula: Mountain Press.

Forman, Richard T., ed. 1979. *Pine Barrens: Ecosystem and landscape.* New York: Academic Press.

Garvey, Edward B. 1971. *Appalachian hiker: Adventure of a lifetime.* Oakton, Va.: Appalachian Books.

Gerrard, Jon M., and Gary R. Bortolotti. 1988. *The bald eagle.* Washington, D.C.: Smithsonian.

Gertler, Edward. 1992. *Garden State canoeing.* Silver Spring, Md.: Seneca Press.

———. 1993. *Keystone canoeing.* Silver Spring, Md.: Seneca Press.

———. 1996. *Maryland and Delaware canoe trails.* Silver Spring, Md.: Seneca Press.

Glassberg, J. 1993. *Butterflies through binoculars: A field guide to butterflies in the Boston, New York, Washington region.* New York: Oxford University Press.

Godfrey, Michael A. 1980. *A Sierra Club naturalist's guide: The Piedmont.* San Francisco: Sierra.

Gosner, Kenneth L. 1978. *A field guide to the Atlantic seashore: Invertebrates and seaweeds of the Atlantic coast from the Bay of Fundy to Cape Hatteras.* Boston: Houghton Mifflin.

———. 1979. *A field guide to the Atlantic seashore.* Boston: Houghton Mifflin.

Grant, P. J. 1986. *Gulls: A guide to identification*. Vermillion, S. Dak.: Buteo.

Green, N. Bayard, and Thomas K. Pauley. 1987. *Amphibians and reptiles in West Virginia*. Pittsburgh: University of Pittsburgh Press.

Greenberg, Russell, and Jamie Reaser. 1995. *Bring back the birds*. Mechanicsburg: Stackpole.

Grimm, William Carey. 1966. *Recognizing native shrubs*. Harrisburg, Pa.: Stackpole.

———. 1968. *Recognizing flowering wild plants*. New York: Hawthorn Books.

Gude, Gilbert. 1984. *Where the Potomac begins*. Cabin John, Md.: Seven Locks Press.

Gurnell, John. 1987. *The natural history of squirrels*. New York: Facts on File.

Hale, Mason E. 1979. *How to know the lichens*. Dubuque, Iowa: Wm. C. Brown.

Harding, John J., and Justin J. Harding. 1980. *Birding the Delaware Valley region*. Philadelphia: Temple University Press.

Headstrom, Richard. 1968. *Nature in miniature*. New York: Knopf.

Heatwole, Henry. 1988. *Guide to the Shenandoah National Park and Skyline Drive*. Luray, Va.: Shenandoah Natural History Association.

Holland, W. J. [1903] 1968. *The moth book*. Reprint, New York: Dover.

Horton, Tom. 1987. *Bay Country: Reflections on the Chesapeake*. Baltimore: Johns Hopkins University Press.

———. 1996. *An island out of time*. New York: Norton.

Horton, Tom, and William M. Eichbaum. 1991. *Turning the tide: Saving the Chesapeake Bay*. Washington, D.C.: Island Press.

Howe, William H. 1975. *The butterflies of North America*. Garden City, N.Y.: Doubleday.

Keatts, Henry. 1995. *Beachcomber's guide from Cape Cod to Cape Hatteras*. Houston: Gulf Publishing.

Kird, P. K., ed. 1979. *The Great Dismal Swamp*. Charlottesville: University of Virginia Press.

Klingel, Gilbert. 1951. *The bay*. Baltimore: Johns Hopkins University Press.

Klots, Alexander B. Field. 1951. *Guide to the butterflies*. Boston: Houghton Mifflin.

Kricher, John C., and Gordon Morrison. 1988. *Eastern forests*. Boston: Houghton Mifflin.

Lambert, Darwin. 1971. *Herbert Hoover's hideaway*. Luray, Va.: Shenandoah Natural History Association.

Lawrence, Susannah. 1984. *The Audubon Society field guide to the natural places of the mid-Atlantic states: Coastal*. New York: Pantheon.

Lawrence, Susannah, and Barbara Gross. 1984. *The Audobon Society field guide to the natural places of the mid-Atlantic states: Inland*. New York: Pantheon.

Leatherman, Stephen P. 1979. *Barrier island handbook*. Washington, D.C.: National Park Service.

Leopold, Aldo. 1949. *A Sand County almanac*. New York: Oxford University.

Lester, Thomas. 1977. *The Pine Barrens of New Jersey*. Trenton, N.J.: Department of Environmental Protection.

Lippson, Alice Jane, and Robert L. Lippson. 1984. *Life in the Chesapeake Bay*. Baltimore: Johns Hopkins University Press.

Lomax, Joseph L., Joan M. Galli, and Anne E. Galli. 1980. *The wildlife of Cape May County, New Jersey*. Pomona, N.J.: Center for Environmental Research, Stockton State College.

Lord, William G. 1965. *The Blue Ridge Parkway guide*. Asheville, N. C.: Stephens Press.

MacKay, Bryan. 1995. *Hiking, cycling, and canoeing in Maryland*. Baltimore: Johns Hopkins University Press.

Martof, Bernard S., William M. Palmer, Joseph R. Bailey, and Julian R. Harrison III. 1980. *Amphibians and reptiles of the Carolinas and Virginia*. Chapel Hill: University of North Carolina Press.

Mazzeo, Peter M. 1981. *Fern and fern allies of Shenandoah National Park*. Luray, Va.: Shenandoah Natural History Association.

McCormick, Jack. 1970. *The Pine Barrens: A preliminary ecological inventory*. Trenton: New Jersey State Museum.

McKnight, Kent H., and Vera B. McKnight. 1987. *A field guide to mushrooms, North America*. Boston: Houghton Mifflin.

McPhee, John. 1967. *The Pine Barrens*. New York: Farrar, Straus & Giroux.

Meanley, Brooke. 1973. *The Great Dismal Swamp*. Washington, D.C.: Audubon Naturalist.

————. 1975. *Birds and marshes of the Chesapeake Bay country*. Cambridge, Md.: Tidewater Publishers.

Miller, Orson K., Jr. 1980. *Mushrooms of North America*. New York: Dutton.

Minichiello, J. Kent and Anthony W. White. 1997. *From Blue Ridge to barrier island: An Audubon naturalist reader*. Baltimore: Johns Hopkins University Press.

Mitchell, Joseph C. 1994. *The reptiles of Virginia*. Washington, D.C.: Smithsonian.

Mittenthal, Suzanne Meyer. 1983. *The Baltimore trail book*. Baltimore: Johns Hopkins University Press.

Mooney, Elizabeth C. 1984. *Country adventures in Maryland, Virginia and West Virginia*. Arlington, Va.: Washington Book Trading.

National Geographic Society. 1983. *Field guide to the birds of North America*. Washington, D.C.: National Geographic.

Newcomb, Lawrence. 1977. *Newcomb's wildflower guide*. Boston: Little, Brown, & Co.

Oberman, Lola. 1986. *The pleasures of watching birds*. New York: Walker & Company.

Opler, Paul. A., and George O. Krizek. 1984. *Butterflies east of the Great Plains: An illustrated natural history*. Baltimore: Johns Hopkins University Press.

Opler, Paul A., and Vichai Malikul. 1992. *A field guide to eastern butterflies*. Boston: Houghton Mifflin.

Page, Lawrence M., and Brooks M. Burr. 1991. *A field guide to freshwater fishes*. Boston: Houghton Mifflin.

Palmer, E. Laurence, and H. Seymour Fowler. 1975. *Fieldbook of natural history*. New York: McGraw-Hill.

Parnes, Robert. 1978. *Canoeing the Jersey Pine Barrens*. Charlotte, N. C.: East Woods Press.

Pasquier, Roger F. 1977. *Watching birds: An introduction to ornithology*. Boston: Houghton Mifflin.

Perry, Bill. 1985. *A Sierra Club naturalist's guide: The middle Atlantic Coast*. San Francisco: Sierra.

Pescatore, John. 1993. *Family bicycling in the Washington-Baltimore area*. McLean, Va.: EPM Publications.

Peterson, Lee. 1978. *A field guide to edible wild plants*. Boston: Houghton Mifflin.

Peterson, Roger Tory. 1980. *A field guide to the birds east of the Rockies*. Boston: Houghton Mifflin.

Peterson, Roger Tory, and Margaret McKenny. 1968. *A field guide to wildflowers of northeastern and north-central North America*. Boston: Houghton Mifflin.

Petrides, George A. 1972. *A field guide to trees and shrubs*. Boston: Houghton Mifflin.

Petry, Loren C. 1968. *The beachcomber's botany*. Chatham, Mass.: Chatham Conservation Foundation.

Phillips, Claude E. 1978. *Wildflowers of Delaware and the Eastern Shore*. Hockessin, Del.: Delaware Nature Education Society.

Pohl, Richard W. 1954. *How to know the grasses*. Dubuque, Iowa: Wm. C. Brown Company.

Pyle, Robert Michael. 1981. *The Audubon Society field guide to North American butterflies*. New York: Knopf.

————. 1984. *Handbook for butterfly watchers*. Boston: Houghton Mifflin.

Radford, Albert E., Harry E. Ahles, and C. Ritchie Bell. 1968. *Manual of the vascular flora of the Carolinas*. Chapel Hill: University of North Carolina Press.

Reed, John C., Jr., Robert S. Sigafoos, and George W. Fisher. 1980. *The river and the rocks: The geological story of Great Falls and the Potomac River gorge*. U.S. Geological Survey Bulletin 1471. Washington, D. C.: U.S. Geological Survey.

Riley, Laura, and William. 1981. *Guide to the national wildlife refuges*. Garden City, N.Y.: Anchor Books.

Robichaud, Beryl, and Murray F. Buell. 1973. *Vegetation of New Jersey*. New Brunswick, N. J.: Rutgers University Press.

Robbins, Chandler S., Bertel Broun, and Herbert Zim. 1983. *Birds of North America*. New York: Golden Press.

Ross, John. 1995. *The Smithsonian guides to natural America: the Atlantic Coast and Blue Ridge*. Washington, D.C.: Smithsonian Books.

Schumann, Walter. 1993. *Handbook of rocks, minerals, and gemstones*. Boston: Houghton Mifflin.

Scott, James A. 1986. *The butterflies of North America*. Stanford, Cal.: Stanford University Press.

Shelton, Napier. 1975. *The nature of Shenandoah*. Washington, D.C.: National Park Service.

Sibley, David. 1993. *The birds of Cape May.* Franklin Lakes, N.J.: New Jersey Audubon Society.

Silberhorn, Gene M. 1976. *Tidal wetland plants of Virginia.* Glouscester Point: Virginia Institute of Marine Science.

Simpson, Bland. 1990. *The Great Dismal.* New York: Henry Holt & Co.

Smith, J. Lawrence. 1982. *The High Alleghenies.* Tornado, W. Va.: Allegheny Vistas.

Smith, James P., Jr. 1981. *A key to the genera of grasses of the conterminous United States.* Eureka, Cal.: Mad River.

Storer, John H. 1953. *The web of life.* New York: Penguin Books.

Strain, Paula M. 1993. *The Blue Hills of Maryland.* Vienna, Va.: Potomac Appalachian.

Strausbaugh, P. D., and Earl L. Core. 1978. *Flora of West Virginia.* Grantsville, W.Va.: Seneca Books.

Taylor, John W. 1992. *Birds of the Chesapeake Bay.* Baltimore: Johns Hopkins University Press.

Teal, John, and Mildred Teal. 1969. *The life and death of the salt marsh.* New York: Ballantine

Teter, Don. 1977. *Goin' up Gandy.* Parsons, W. Va.: McClain.

Tiner, Ralph W., Jr. 1987. *A field guide to coastal wetland plants of the northeastern United States.* Amherst: University of Massachusetts Press.

Tiner, Ralph W., Jr. 1988. *Field guide to nontidal wetland identification.* Annapolis: Maryland Department of Natural Resources and U.S. Fish and Wildlife Service.

Tyning, Thomas F. A 1990. *Guide to amphibians and reptiles.* Boston: Little, Brown & Co.

Ursin, Michael J. 1972. *Life in and around the salt marshes.* New York: Thomas Y. Crowell.

Vankat, John L. 1979. *Natural vegetation of North America.* New York: Wiley.

Vokes, H. E., and J. Edwards, Jr. 1974. *Geography and geology of Maryland.* Maryland Geological Survey Bulletin 19. Annapolis: Maryland Department of Natural Resources.

Warner, William W. 1976. *Beautiful swimmers.* Boston: Little, Brown and Co.

Webster, W. D., J. F. Parnell, and W. C. Biggs. 1985. *Mammals of the Carolinas, Virginia, and Maryland.* Chapel Hill: University of North Carolina Press.

Weidensaul, Scott. 1992. *Seasonal guide to the natural year: mid-Atlantic.* Golden, Colo.: Fulcrum Publishing.

Weiss, Raymond B. 1969. *A scenic guide to the Monongahela National Forest.* Parsons, W.Va.: McClain Printing Company.

Wennerstrom, Jack. 1995. *Soldiers delight journal.* Pittsburgh: University of Pittsburgh Press.

White, Christopher P. 1989. *Chesapeake Bay, nature of the estuary.* Centreville, Md.: Tidewater Publishers.

White, Richard E. 1983. *A field guide to beetles.* Boston: Houghton Mifflin.

Wilds, Claudia. 1983. *Finding birds in the national capital area.* Washington, D.C.: Smithsonian.

Williams, John Page, Jr. 1993. *Chesapeake almanac.* Centreville, Md.: Tidewater Publishers.

Wofford, B. Eugene. 1989. *Guide to the vascular plants of the Blue Ridge.* Athens: University of Georgia Press.

Wright, Amy Bartlett. 1993. *Peterson first guides: Caterpillars.* Boston: Houghton Mifflin.

Index

Aberdeen Proving Ground, 163

Alexanders, Golden, 135

Alexandria, 84, 87, 137, 218, 221

Alga, 16, 60

Allegheny Front, 107, 108

Allegheny Mountain, 110

Allegheny Plateau, 2, 3, 6, 7, 8, 10, 30, 108, 109, 110, 113, 143, 145

Alpena, 111

Alumroot, 13

Amphibolite, 136

Anacostia River, 208, 209, 210

Anemone: Rue, 13, 45; Wood, 149

Annapolis Rocks, 142

Anole, Green, 173

Ant, 32, 48

Antietam Creek, 128

Antietam National Battlefield, 128

Antlion, 33, 63

Aphid, 32

Appalachian Mountains, 29, 76, 144, 146

Appalachian Trail, 6, 24, 130, 141–156, 142 (map)

Apple, May, 12

Arbutus, Trailing, 135, 210

Arrowhead, Broad-leaved, 12

Arum, Arrow, 11, 131

Ash: 9, 130; Green, 14; Pumpkin, 173; Water, 173; White, 14

Aspen, Bigtooth, 14

Assawoman Bay, 189

Assateague Island, 82, 98, 179, 190, 191, 192

Asters: 9, 222; Serpentine, 215

Atlantic Ocean, 22, 25, 50, 71, 81, 88, 97, 164, 177, 178–194, 180 (map)

Audubon Naturalist Society, 18, 35, 52, 207, 208

Azaleas: Rosy Wild, 65, 108; Wild, 14, 54, 209

Back Bay National Wildlife Refuge, 194

Back River, 164

Back River Sewage Treatment Plant, 88, 164

Baldcypress, 11, 14, 172, 173, 174, 175, 193, 214

Balm, Bee, 64

Baltimore, 6, 13, 28, 38, 63, 88, 154, 163, 164, 170, 183, 195, 197, 210, 212, 214, 216, 220

Basalt, 130

Bass: 130, 137; Largemouth, 31; Black Sea, 31; Smallmouth, 31; Striped, 31

Basswood, 9, 10, 14, 147

Bats: 20, 22, 66, 68, 86, 89, 129; Big Brown, 22; Evening, 22; Indiana, 22, 129; Red, 22; Townsend's Big-eared, 22

Bath Alum, 145

Bath County, 145, 146

Batsto River, 184

Battle Creek, 214

Bays: Red, 175; Sweet, 14

Bayberry, 12

Bears: Black, 19, 20, 85, 86, 108, 116, 147, 173; Wooly, 34

Bear Rocks, 65, 108

Beauties: Meadow, 13; Spring, 12, 44, 45, 132, 217

Beaver, 10, 21, 104, 108, 111, 116, 128, 131, 139, 175, 200, 204, 210, 214, 218

Beaverdam Creek, 209

Bees: 32, 48; Carpenter, 63

Beech, American, 9, 10, 11, 14, 61, 110, 138, 206

Beetles: 32, 34, 48, 66, 70, 72; June, 63; Long-horned, 32, 128; Scarab, 32

Beetree Run, 216

Belle Haven Picnic Area, 90, 137

Bellwort, Perfoliate, 45

Beltsville Agricultural Research Center, 209

Bemis, 110, 111

Bennett Creek, 200, 201, 202

Bestpitch, 168

Bethany, 189

Big Blue Trail, 150, 151, 152

Big Pool, 127, 128

Big Devil's Stairs, 148

Big Savage Mountain, 123

Big Schloss, 151

Billy Goat Trail, 135, 136

Bindweed, Shale, 10

Birches: 9, 116; Black, 10, 14, 110; River, 14, 173; Yellow, 10, 14, 110

Bird feeders, 89, 99, 100

Bird River, 216

Birdathon, 52, 56

Bishops Head, 168

Bison, 20

Bitterns: American, 219; Least, 131, 219

Blackberry, 200, 216

Blackbirds: Red-winged, 138; Rusty, 47

Black Hill Regional Park, 84, 203, 204

Black Marsh, 163

Black Rock, 142

Blackwater Falls, 111, 112

Blackwater National Wildlife Refuge, 60, 71, 84, 91, 166, 167, 168

Blackwater River, 111, 112

Bladensburg, 209

Blister Swamp, 113

Blockhouse Point, 132

Bloodroot, 13

Bluebell, Virginia, 12, 45, 132, 137, 216, 217

Blueberry, 65, 66, 108, 109, 209

Bluebird, Eastern, 47, 90, 100, 127, 131, 163, 209, 213, 217, 219, 222

Bluefish, 31, 189

Blue Mountain, 153, 155

Blue Ridge, 6, 76, 77, 129, 143, 145, 146, 147, 148, 150, 153, 155, 222

Blue Ridge Parkway, 144, 146

Blues: Dusky, 48; Eastern Tailed, 68; Silvery, 33, 48

Bluestems: 215; Little, 217

Bobcat, 17, 19, 20, 91, 147, 152, 173

Bobolink, 52, 109, 112

Bolete, 16

Bombay Hook, 51, 84, 186, 188, 189

Boonsboro, 155

Bother Knob, 145

Boundary Bridge, 206

Bowie, 212

Brant, 75, 189

Breeches, Dutchman's, 45

Brighton Dam, 211

Broad Bay, 171

Broad Run, 104

Broadkill Beach, 51

Broadkill Marsh, 98, 186, 187

Buck Mountain, 105

Buckeye, 74, 82, 184

Buena Vista, 143

Bufflehead, 46, 84, 97, 140, 165, 181

Bugs: 32, 70; Assassin, 32; Boxelder, 32; Milkweed, 32; Shield, 60, 63; Stink, 32

Bullfrog, 17, 27, 53, 60, 131, 211

Bull Run, 104, 217

Bull Run Mountain, 6, 217

Bull Run–Occoquan Trail, 217, 218

Bull Run Regional Park, 217

Bulrushes: Great, 11; Saltmarsh, 11

Bunchberry, 13

Bunting, Indigo, 208, 216, 222

Bunting, Snow, 97, 165, 181, 185

Burnet, Canada, 123

Butter, Witch's, 16

Buttercup, Swamp, 45

Buttonbush, 11, 213, 219
Buzzard Rocks, 151

Cabbage, Skunk, 12, 45, 88, 92, 95
Cabin Mountain, 66
Cacapon River, 105, 127
Caddisfly, 32, 34, 73
Caldwell, 114
Caledon Natural Area, 139
Calfpasture River, 145
Calvert Cliffs State Park, 170
Camp David, 154
Camp Hoover, 147
Canaan Mountain, 113
Canaan Valley, 111, 112
Canvasback, 83, 90, 162
Cap, Inky, 16
Cape Charles, 81, 177, 181, 193
Cape Henlopen, 98, 185, 186, 189
Cape Henry, 193
Cape May, 81, 181, 183, 185
Capon Springs Resort, 104
Carderock, 134
Cardinal, Northern, 90, 93, 99, 204
Carlisle, 153, 156
Cass, 114
Catfish, 31
Catoctin Mountain, 130, 154
Catoctin Mountain Park, 154
Catonsville, 215
Cattails: Broad-leaved, 11, 131; Narrow-
    leaved, 11
Cave, 65, 104, 109, 129, 150
Cedars: Atlantic White, 173, 184; Red, 147,
    201, 220
Chance, 176
Chanterelle, Golden, 16, 108
Chat, Yellow-bellied, 176, 201, 222
Cheat Bridge, 110, 113
Cheat Mountain, 113
Cheat River, 109, 110, 111, 113
Checkerspots: Baltimore, 200; Harris', 112,
    123; Silvery, 131
Cherry River, 117
Cherry trees: 9, 113, 147, 153; Black, 10, 12,
    14, 110
Chesapeake and Ohio (C & O) Canal, 46,
    47, 60, 71, 89, 90, 124, 125, 127, 128,

130, 131, 132, 134, 135, 136, 141, 154,
    204, 205
Chesapeake Bay, 21, 22, 31, 35, 50, 71, 81,
    83, 84, 86, 87, 91, 137, 138, 157–177,
    160 (map), 210, 213, 214, 215, 216
Chesapeake Bay Bridge, 84, 164, 165
Chesapeake Bay Bridge and Tunnel, 193
Chester River, 165
Chestertown, 165
Chestnut, American, 10, 11, 147, 211
Chevy Chase, 67
Chickadees: Black-capped, 111; Carolina,
    77, 90, 92, 93, 99
Chickweed: Common, 96; Star, 45
Chimney Rock, 153
Chimney Tops, 107
Chincoteague National Wildlife Refuge,
    81, 98, 191, 192
Chinquapin, 14
Chipmunk, Eastern, 20, 21, 86, 204
Choptank River, 166, 168
Christmas Bird Count, 86, 88, 90, 203
Chub, Creek, 31
Chuck-will's-widow, 56, 169
Cicadas: 33, 60; Annual, 62; Periodical, 37,
    62, 63
Cinquefoil, Rough-fruited, 13
Civil War, 104, 127, 128, 129, 132, 154, 200,
    217
Claggett Farm, 212
Clam, 170
Clark's Crossing Park, 222
Cliffbrake, Purple, 15, 107, 128
Clintonia, White, 123
Cloak, Mourning, 33, 48, 92, 96
Clopper Lake, 203
Close to home, 198–199 (map)
Clover, Kates Mountain, 10, 105, 126
Clubmoss, 62
Coal, 35, 105, 119
Coastal Plain, 2, 6, 11, 24, 28, 35, 70, 83, 86,
    138, 158, 163, 167
Cohosh, Blue, 45, 147
College Park, 210
Colonial Beach, 139
Columbia, 212
Columbine, 13
Comma, Eastern, 33, 48, 96

Conn Island, 88

Conowingo Dam, 88, 162

Coot, American, 93

Cooter, River, 24

Copper, American, 123, 127

Copperhead, 23, 54, 105

Coralroot, Spotted, 13

Corapeake Ditch, 175

Cordgrass: Big, 11, 159, 168; Saltmarsh, 11, 91, 159

Cormorants: Double-crested, 83, 137, 171; Great, 171, 185

Corn, Squirrel, 45

Cottonmouth, 24, 172

Coyote, 20

Crabs: 161, 166, 176, 179, 182, 184, 190, 192; Horseshoe, 51, 52, 182, 186, 187, 188

Crabtree Falls, 145

Crampton's Gap, 154

Cranberry, 13, 184

Cranberry Backcountry, 116

Cranberry Botanical Area, 117

Cranberry Glades, 117

Cranberry River, 116

Cranberry Visitor Center, 115

Cranberry Wilderness, 116

Crane, Whooping, 212

Crayfish, 34

Creepers: Brown, 74, 90, 93; Trumpet, 14, 62; Virginia, 14, 62

Crescent, Pearl, 33, 48, 68

Cress: Rock, 45; Winter, 75

Crickets: 32, 63, 66, 67, 70, 73, 78; Field, 78; Snowy Tree, 78

Croaker, Atlantic, 31

Crocheron, 168

Crossbills: Red, 90, 115, 208; White-winged, 207

Crossvine, 15, 173, 174

Crow, American, 76, 164

Cuckoo, Yellow-billed, 174, 176

Cumberland, 124, 125, 134, 152

Cunningham Falls State Park, 91, 154

Cup, Scarlet, 16

Dace: Black-nosed, 31; Rosy-sided, 31

Damselfly, 34, 49, 191

Dandelion, 47, 96

Dans Mountain, 124

Darters: Fantail, 31; Shield, 31; Tessellated, 31

Dead-nettles: Purple, 96

Deal Island, 176

Dean's Gap, 153

Deer, White-tailed, 19, 20, 78, 79, 89, 91, 116, 138, 147, 151, 187, 206, 207

Delaware, 21, 175, 185, 186, 189, 193

Delaware Bay, 50, 51, 182, 183, 185, 186, 187, 188

Delaware Wildlands, 175

Delmarva Peninsula, 181, 192

Devil, Hickory Horn, 34

Dewberry, 66

Dewey Beach, 189

Diersson Wildlife Management Area, 46

Difficult Run, 135

Dobsonfly, 33, 63

Dock, Water, 11

Dogwood, Flowering, 14, 61

Dolly Sods, 65, 107, 108

Dolphin, Atlantic Bottle-nosed, 22, 171, 190

Dorchester County, 166, 168

Douglas, Justice William O., 125

Douthat State Park, 146

Dove, Mourning, 99

Dover, 187, 188

Dragonfly, 17, 32, 34, 49, 60, 62, 73, 74, 131, 191, 204

Drums: Black, 31; Red, 31

Ducks: Harlequin, 189; Ring-necked, 46, 131; Ruddy, 83, 90; Wood, 46, 131, 204, 205

Duckweed, 60

Dune, 12, 97, 98, 171, 172, 179, 191, 193, 194

Dunlin, 50

Duskywing, Juvenal's, 48, 135

Dyke Marsh, 90, 137, 138, 220

Eagles: Bald, 80, 87, 98, 137, 139, 162, 163, 166, 169, 171, 212; Golden, 76, 80, 87, 166

Earthstar, 16

Earthworm, 17

Eastern Neck National Wildlife Refuge, 84, 164, 165

Eastern Shore, 14, 56, 83, 84, 87, 161, 175, 176
Eastern Shore Birding Festival, 193
Eastern Shore of Virginia National Wildlife
    Refuge, 177, 193
Eft, Red, 60
Egret, Reddish, 183, 188
Eider, 189
Elder, Box, 10
Elfin, 48, 176
Elfin, Henry's, 33
Elizabeth Furnace, 151
Elk, 20
Elk Neck, 162, 163
Elk River, 163
Elkins, 113
Ellicott City, 215
Elliott Island, 28, 56, 91, 168, 169, 170
Elms: 96, 130; Slippery, 10
Elm Island, 132
Emperor, Tawny, 132
Evening-Primrose, Shale Barren, 126
Eye, Northern Pearly, 135

Fairfax Stone, 54, 119
Falcon, Peregrine, 51, 80, 82, 86, 98, 134,
    164, 165, 171, 185, 187, 190, 193
Fall Line, 6, 11, 132, 137, 205
Falls of Hill Creek, 117
False Cape State Park, 194
Fameflower, 215
Fennel, Dog, 12
Ferns: 15; Christmas, 75, 90; Cinnamon, 123;
    Curly-grass, 184; Walking, 10, 128
Fetterbush, 175
Finches: House, 99; Purple, 77, 90, 100, 115
Finger's, Dead Man's, 16
Finzel Swamp, 123
Fir, Balsam, 10, 13, 110, 113
Firefly, 32, 62, 66, 67, 208
Fisherman's Island National Wildlife Refuge,
    177
Fishfly, 33
Fishing Bay Wildlife Management Area, 168
Flag, Sweet, 131
Flicker, Northern, 90
Flies: 32, 48, 59, 62; Black, 32; Deer, 32;
    Horse, 32

Floodplain, 11, 45, 47, 54, 122, 127, 132, 133,
    135, 136, 158, 159, 217
Flounder, 31
Flower, Cardinal, 13, 64, 213
Flycatchers: Acadian, 59, 153, 174; Alder,
    112, 115, 124; Fork-tailed, 171; Least, 108;
    Willow, 131, 201; Yellow-bellied, 52
Foamflower, 13
Forsythe National Wildlife Refuge, 183
Fort Frederick, 127, 128
Fort McHenry, 164
Fort Meade, 212
Fort Smallwood, 81
Fort Valley, 151
Fossil, 35, 170
Fourth Street Flats, 189
Foxes: 78, 91, 131, 133; Gray, 20, 167, 187;
    Red, 20, 130, 165, 186, 207
Frederick, 154, 200
Fredericksburg, 6
Freezeland Flat, 145
Friendship Landing, 138
Friends of Huntley Meadows, 219
Fringe-tree, 135
Fritillaries: Aphrodite, 127; Atlantis, 112, 123;
    Diana, 146; Great Spangled, 68, 131, 201;
    Meadow, 162; Silver-bordered, 111
Frogs: Carpenter, 28; Green, 27, 55, 60, 131;
    Pickerel, 27, 55, 131, 211; Wood, 25, 26,
    55, 96, 131, 207
Fungus, 15, 48, 75

Gallberry, 175
Gambrill State Park, 154
Gandy Creek, 109, 110
Gannet, Northern, 97, 165, 181, 190, 192
Gathland State Park, 154, 155
Gaudineer Knob, 113
Geese: Blue, 166; Canada, 46, 55, 79, 83, 84,
    90, 131, 137, 165, 166, 167, 186, 213, 219;
    Ross', 203; Snow, 77, 83, 84, 98, 166,
    184, 186
Gentian, Fringed, 215
George Washington Memorial Parkway, 136
George Washington National Forest, 105,
    144, 145, 146, 150
Georgetown, 134, 205

Georgetown Reservoir, 88

Geranium, Wild, 149

Ginger, Wild, 12

Ginseng, Dwarf, 45

Glasswort, 181

Gnat, 62

Goldeneye, Common, 84, 140, 165, 181

Goldenrod: 9, 123, 128, 217, 222; Seaside, 12, 82, 181

Goldfinch, American, 99, 100

Goldthread, 13

Goshawk, Northern, 76, 80

Granite, 136

Grape, Wild, 14, 62

Grass: 9; Beach, 12, 181; Indian, 217

Grasshopper, 32, 78

Gravelly Springs Gap, 147

Graysonville, 166

Great Dismal Swamp, 15, 24, 28, 172, 173, 174

Great Falls, 44, 81, 88, 132, 134, 135, 136

Great Marsh, 84

Great North Mountain, 104, 151, 152

Great Valley, 6, 128, 129, 153

Grebes: 83; Eared, 165; Horned, 83, 97, 140, 165, 181; Red-necked, 205

Greenbelt Park, 209

Green Ridge State Forest, 125, 126, 127, 152

Greenbrier, 12

Greenbrier River, 113, 114

Greenstone, 130, 154

Grosbeaks: Blue, 169; Evening, 90, 100; Rose-breasted, 54, 123, 148

Groundsel, Everlasting, 105

Grouse, Ruffed, 91, 123, 145

Gulls: Black-headed, 88, 192; Bonaparte's, 190; Glaucous, 88; Great Black-backed, 88, 181; Herring, 88, 137, 164; Iceland, 88, 137; Laughing, 88; Lesser Black-backed, 88, 137; Little, 88; Ring-billed, 88, 137, 164; Ross', 88, 164, 189; Thayer's, 88; Yellow-legged, 88

Gum, Black, 14, 134, 147

Gum, Swamp Black, 173

Gum, Tupelo, 173

Gunpowder Falls, 215, 216

Gunpowder River, 215

Hackberry (butterfly), 68, 132

Hackberry (tree), 14, 132

Hagerstown Valley, 6, 128, 154

Hairstreak: Coral, 68, 201; Great Purple, 174, 194; Olive, 201; White-M, 140

Hampton Roads, 171

Hancock, 118, 127, 152

Harbinger-of-spring, 45, 133

Hardscrabble Knob, 145

Harman, 113

Harperella, 127

Harpers Ferry, 125, 129, 148, 153, 154

Harrier, Northern, 80, 87, 91, 98, 138, 166, 187

Harrisburg, 161

Harrisonburg, 145

Harry Diamond Lab, 218

Hart-Miller Island, 163

Harvester, 131

Havre de Grace, 158, 162

Hawks: Broad-winged, 71, 80; Cooper's, 80, 81, 100; Red-shouldered, 80, 204, 217; Red-tailed, 76, 80, 98, 166, 217; Rough-legged, 75, 80, 87, 98, 168; Sharp-shinned, 80, 81, 98, 100, 156

Hawk Campground, 104

Hawk Mountain, 80, 81, 156

Hawkweed, 119

Hay, Saltmeadow, 11, 159

Hazel, Witch, 12, 76, 77

Heart, Bleeding, 109

Heather, Beach, 12

Hellgrammite, 33

Hemlock, Eastern, 10, 110, 145, 147, 217

Hemlock Overlook Regional Park, 217

Henbit, 96

Henrys Crossroads, 168

Hepatica, 44, 45

Herndon, 221, 222

Herons: Great Blue, 93, 131, 139, 167, 189; Green, 59, 131, 204, 219

Hibiscus, Marsh, 11

Hickories: 9, 10, 11, 88, 90, 147, 153, 195, 211, 216; Mockernut, 14; Pignut, 14; Shagbark, 14

Highland Scenic Highway, 115, 116

Highland Trace, 103–117, 106 (map)

Holly, American, 11, 12, 14, 61, 90, 91, 138
Honeysuckle, Japanese, 222
Hooper Island, 168
Hornwort, 15
Horsehead Sanctuary, 165
Horsetail, 15, 135
Huckleberry, 66
Hughes Hollow, 44, 47, 60, 86, 96, 130, 131
Hummingbird, Ruby-throated, 62, 64, 115
Hunting, 77, 123, 130, 149, 151, 175, 200
Hunting Quarter Road, 131, 132
Huntley Meadows, 60, 71, 218, 219, 220
Hydrangea, Climbing, 15

Ibis, White, 183, 188
Igneous, 3, 136
Indian River Inlet, 189
Indigo, Blue False, 135
Ironweed, New York, 13, 222
Ivy, Poison, 14, 38, 62

Jack-in-the-pulpit, 12
Jack-o-lantern, 16
James River, 143, 158, 171
James River Face Wilderness, 144
Jericho Ditch, 174
Jessamine, 15, 174
Jug Bay, 84, 213
Jug Bay Wetlands Sanctuary, 213
Junco, Dark-eyed, 99
Juniper, Virginia, 9, 14

Katydid, 32, 63, 66, 70, 78
Kenilworth Aquatic Garden, 209
Kent Island, 165, 166
Kent Narrows, 166
Kestrel, American, 80, 82, 86, 130
Killifish: Banded, 31; Rainwater, 31
Kingfisher, Belted, 59
Kinglets: Golden-crowned, 72, 90, 93; Ruby-crowned, 72, 77
Kingsley Schoolhouse, 201
Kingsnakes: Eastern, 149; Scarlet, 24
Kiptopeke State Park, 177, 193
Kittatinny Ridge, 80, 156
Kitts Hummock, 51, 186
Kitzmiller, 119

Klein, Ivan, 65
Knot, Red, 51

Lace, Queen Anne's, 13
Lacewing, 32, 33, 63, 66
Ladybug, 32, 62, 208
Ladyslippers: Pink, 182; Yellow, 13, 54, 136, 147, 149
Lake Drummond, 174, 175
Lake Frank, 208
Lake Needwood, 208
Lamprophyre, 136
Lancaster, 161
Laneville, 108
La Plata, 138
Laurels: 212; Mountain, 10, 14, 54, 62, 65, 123, 153, 203, 209; Sheep, 13, 175
Laurel Fork Campground, 109
Laurel Fork Wilderness, 111
Laurel Prong, 147
Lavender, Sea, 82, 181
Leafhopper, 32, 60, 64
Leatherflower, Whitehaired, 10
Lee, Robert E., 171
Leek, Wild, 133
Leopold, Aldo, 56
Lettuce, Wild, 13
Leucothoe, 175
Lewes, 185
Lewis Spring Falls, 148
Liberty Reservoir, 214
Lichen, 15, 16, 17
Lilies: Calla, 123; Canada, 149; Water, 60
Lilypons Water Garden, 60, 202
Limestone, 10, 15, 65, 105, 107, 109, 128, 129, 146, 150
Limpkin, 183
Linden, 54, 148
Lion, Mountain, 20
Little Bennett Park, 200, 201
Little Blackwater River, 112
Little Creek, 51, 186, 187, 188
Little Falls, 216
Little Gunpowder Falls, 215, 216
Little Hunting Creek, 90
Little Passage Creek, 151
Little Patuxent River, 212

Little Seneca Creek, 203
Little Seneca Lake, 84, 203, 204
Liverwort, 15
Lizards: Eastern Fence, 25; Slender Glass, 25
Lizard's-tail, 13, 60, 219
Loch Raven Reservoir, 216
Locust, Black, 9, 14, 153
Loons: 83, 93, 189; Common, 97, 140, 165,
    170, 181, 202; Red-throated, 97, 171,
    181, 185
Lost River, 105, 106
Lost River State Park, 105
Loudoun Heights, 153
Luke, 122, 124
Lyme Disease, 38

Magnolias: 14; Southern, 173; Swamp, 173,
    175
Mallows: 138; Crimson-eyed Rose, 13;
    Rose, 60
Manassas, 104, 217
Manassas National Battlefield Park, 217
Mantid, 32, 60, 62, 208
Maples: 9, 61, 92, 108, 116, 130, 147, 153;
    Red, 11, 12, 14, 44, 47, 77, 88, 96, 134,
    173, 209, 214; Silver, 10, 14, 44, 162, 217;
    Sugar, 10, 14
Marble, Olympia, 48
Mark, Question, 33, 48, 96
Marlinton, 114, 115
Maryland: 21, 44, 50, 54, 56, 65, 71, 81, 90,
    118, 119, 122, 125, 127, 128, 129, 130, 134,
    135, 138, 139, 142, 148, 149, 150, 152, 153,
    154, 155, 158, 162, 166, 170, 175, 179, 189,
    190, 191, 193, 195, 200, 203, 206, 207, 209,
    210, 214
Maryland Heights, 153
Mason-Dixon Line, 118, 127
Mason Neck, 52, 71, 84, 86, 87, 90, 91, 138,
    217, 218
Massanutten Mountains, 150, 151
Massanutten Peak, 151
Mather Gorge, 134, 135, 136
Mayflower, Canada, 13, 123
Mayfly, 32, 34, 49, 54, 73
McKee-Beshers Wildlife Management Area,
    47, 130, 131

Meadowlark, Eastern, 189
Meadowside Nature Center, 208
Menhaden, Atlantic, 31
Mergansers: Common, 46, 84, 132, 162;
    Red-breasted, 97, 181, 189
Merkle Wildlife Sanctuary, 84, 213
Merlin, 80, 82
Mermaids, False, 137
Metagraywacke, 135, 136
Metalmark, Northern, 127
Metamorphic, 3, 130, 132, 135, 136
Mice: 78, 79, 91; Deer, 21; Harvest, 21;
    House, 21; White-footed, 21
Michaux State Forest, 141
Middle Mountain Road, 110, 111
Middle Patuxent River, 212
Midge, 49
Migration Fallout, 172
Milburn Landing, 175
Milkweed: Common, 68, 131, 146; Green,
    217; Red, 13; Swamp, 13, 213, 222
Mill Prong, 147
Millet, Walter's, 11
Mink, 20
Minnows: Fathead, 31; Sheepshead, 31
Mockingbird, Northern, 17, 47
Monadnock, 6, 201
Monarch, 64, 65, 68, 71, 74, 82, 179, 193
Monocacy National Battlefield, 200
Monocacy Natural Resources Management
    Area, 200
Monocacy River, 200, 202
Monongahela National Forest, 107, 109,
    110, 113
Monument Knob, 81, 155
Moorefield, 105, 106, 107
Morel, 16, 48, 211
Moss: 15; Spanish, 172; Sphagnum, 117, 175
Moths: 32, 34, 66; Luna, 34; Sphinx, 34, 208
Mount Vernon, 87
Mount Vernon Parkway, 88, 90, 137, 222
Mountain Lock, 128
Mullica River, 184
Mummichog, 31
Mushrooms: 15, 16, 48, 75, 154; Meadow, 16
Muskrat, 22, 165, 167, 176, 187, 218

Myotis, Little Brown, 22
Myrtle, Wax, 12, 98, 181, 191

Nanjemoy Creek, 138, 139
Nassawango Creek, 176
National Arboretum, 209
National Audubon Society, 86
National Park Service, 129, 144, 171, 191, 205
National Pike, 124
National Zoo, 205
Nature Conservancy, 105, 107, 123, 139, 173, 176, 192
Needlerush, Black, 11
Nettle, Stinging, 38
New Germany State Park, 123
New Jersey, 22, 181, 183, 186
Newport News, 171
Newt, Red-spotted, 60, 111
New York: 161
Night-Heron, Black-crowned, 206
Norfolk, 158, 171
North Carolina, 144, 194
North Central Railroad, 216
North Fork Mountain, 107
North Lookout, 80
North Point State Park, 163
Northern Neck, 171
Nuthatches: Brown-headed, 139, 168; Red-breasted, 90, 100; White-breasted, 90, 93
Nutria, 22, 167

Oaks: 9, 10, 11, 61, 76, 88, 90, 113, 125, 130, 153, 154, 171, 206, 209, 211, 216, 220; Bear, 14; Blackjack, 14, 184; Chestnut, 14, 147, 151; Laurel, 173; Pin, 14; Post, 12, 14, 135, 184; Red, 14, 147, 151; Scarlet, 12; Southern Red, 14; Swamp Chestnut, 173; Water, 173; White, 11, 14, 139, 168; Willow, 14, 134
Occoquan Reservoir, 217
Occoquan River, 138, 216, 217
Ocean City, 82, 88, 179, 189, 190, 191
Ocean City Inlet, 190
Old Angler's Inn, 134
Oldsquaw, 84, 97, 140, 165, 181
Onion, Shale, 10, 105
Opossum, 22, 206

Orange-tip, Falcate, 33, 48
Orchids: Cranefly, 13; Round-leaved, 13; Snake-mouth, 117; Bog, 184
Orchis, Showy, 149
Orioles: Baltimore, 200, 211; Orchard, 176
Orthopterans, 32, 70, 78
Osprey, 46, 80, 166, 187, 204, 213
Oswego River, 184
Otter Creek, 111
Otter, River, 20, 214
Ovenbird, 59, 145, 174
Owens Creek, 154
Owls: Barn, 56; Barred, 49, 79, 131, 138, 175; Great Horned, 59, 66, 79, 86, 92, 94, 98, 109, 187, 206; Northern Saw-whet, 108, 163, 191; Short-eared, 87, 91, 170, 187; Snowy, 98, 164, 185
Oyster, 161, 170, 176
Oystercatcher, American, 189

Parakeet, Carolina, 9
Parsons, 111
Partridgeberry, 90
Parula, Northern, 49
Passage Creek, 151
Patapsco River, 164, 214, 216
Patapsco River, North Branch, 214
Patapsco River, South Branch, 214, 215
Patapsco Valley State Park, 214, 215
Patuxent River, 84, 170, 210, 212
Patuxent River Park, 213
Patuxent River State Park, 211
Patuxent Wildlife Research Center, 212
Paw Paw, 127
Pawpaw, 14
Pawpaw Bends, 125
Peaks of Otter, 30, 144
Peel, Orange, 16
Peeper, Spring, 25, 26, 28, 54, 55, 97, 131, 207
Pelican, Brown, 171, 189
Pennsylvania, 65, 80, 81, 118, 141, 149, 150, 153, 154, 155, 161, 216
Pennyfield Lock, 44, 46, 47, 132, 134
Perch, White, 31
Perch, Yellow, 31
Perryville, 162

Persimmon, 14
Petersburg, 107, 108, 110
Petunia, Wild, 217
Phacelia, Colville's, 133
Pheasant, Ring-necked, 189
Phloxes: Blue, 45; Moss, 13; Swordleaf, 10
Phoebe, Eastern, 47, 49, 104, 219
Pickerelweed, 11, 13, 131, 213
Piedmont, 2, 6, 8, 11, 24, 35, 44, 45, 46, 84,
    90, 130, 133, 136, 153, 161, 201, 203, 214,
    215, 217, 222
Pigeon, Passenger, 9
Pimpernel, Mountain, 126
Pines: Loblolly, 9, 11, 12, 14, 139, 168, 191;
    Pitch, 14, 184; Pond, 175; Table Mountain,
    14, 147, 151; Virginia (Scrub), 9, 14, 135,
    147, 209, 220; White, 14
Pine Barrens, 184, 185
Piney Run, 215
Pink, Fire, 13, 107
Pink, Grass, 117
Pinnacle Ridge, 145
Pintail, Northern, 90
Pipevine, 115
Pipistrelle, Eastern, 22
Pipit, American, 47
Plant, Pitcher, 13, 184
Plantain, Rattlesnake, 13
Planthoppers, 32
Plover, Piping, 185
Pocahontas County, 114
Pocomoke River, 28, 175, 176
Pocosin, 173, 175
Pogonia, Large Whorled, 136
Pogonia, Rose, 13
Point Lookout, 54, 81, 139, 140, 170
Point of Rocks, 45, 130, 153
Polygala, Fringed, 13
Polypody, Rock, 15
Poole's Store, 204
Pope's Creek, 171
Pork, Sea, 192
Port Mahon, 187
Potomac River: 44, 45, 46, 47, 54, 83, 84, 87,
    88, 89, 90, 103, 109, 110, 118–140, 148, 153,
    158, 170, 171, 197, 203, 204, 205, 209, 216,
    217, 220, 120–121 (map); North Branch,

119, 127; North Fork of South Branch,
    108; South Branch, 106, 107, 127
Prettyboy Reservoir, 216
Priest Mountain, 145
Prime Hook, 186
Primrose, Evening, 13
Prince William Forest Park, 220
Puffball, 16
Purcellville, 221
Purple, Red-spotted, 68
Puttyroot, 136

Quartzite, 108, 201
Quillwort, 15

Rabbits: Eastern Cottontail, 22, 91, 130;
    Marsh, 173
Raccoon, 20, 79, 133, 179, 192, 206, 214, 218
Rails: Black, 56, 57, 168, 169; Clapper, 91,
    169, 187; King, 169, 219; Virginia, 56, 219
Railroad Ditch, 175
Ramps, 133, 137
Ramsey's Draft, 145
Rappahannock River, 158, 171
Rats: Black, 21; Marsh Rice, 21; Norway, 21
Rattlesnakes: 17, 23, 54, 115; Canebrake, 24,
    173; Timber, 23, 24, 145
Raven, Common, 76, 77, 115, 202
Reading, 156
Recluse, Brown, 38
Red Creek, 108, 111
Redbud, 14
Reddish Knob, 145
Redpoll, Common, 90, 100
Redstart, American, 49
Reed, Common, 12
Rhododendron, 10, 14, 65, 113, 116, 123
Richmond, 6
Richwood, 103, 114, 117
Riverbend Park, 136
Riverside Picnic Area, 88, 90
Roach, 32
Roanoke, 30
Robertson Mountain, 72
Robin, American, 47
Rock Creek, 205, 207, 208
Rock Creek Park, 205, 206

Rock Creek Regional Park, 208
Rocket, Sea, 12
Rockfish Gap, 155
Rocky Gap, 152
Rocky Gorge, 211
Rocky Mountain Spotted Fever, 38
Rose, Multiflora, 222
Rose River Falls, 148
Ruff, 51, 188
Rushes: Needle, 159; Soft, 11
Russula, 16
Rust, Cedar Apple, 16

Sachem, 70
Salamanders: Appalachian Seal, 108; Cheat
    Mountain, 29, 113; Green, 115; Jefferson,
    29; Long-tailed, 29, 201; Marbled, 29;
    Mud, 29; Northern Dusky, 29, 60; Peaks
    of Otter, 30, 144; Red, 29; Red-backed,
    29, 54, 55, 133; Shenandoah, 30; Slimy, 29,
    54; Spotted, 28, 29, 54, 55, 207; Spring, 29,
    149; Three-lined, 29; Tiger, 29; Two-
    lined, 29, 60, 149, 211; Wehrle's, 29, 113
Saltgrass, 11
Sanderling, 82, 181
Sandpipers: 50; Curlew, 51, 188; Purple, 189
Sandstone, 10, 15, 105, 125, 135, 151, 204
Sandwort, Rock, 128
Sandy Point State Park, 81, 84, 164, 165
Sapsucker, Yellow-bellied, 77, 90
Sassafras, 14
Sassafras River, 163
Satyr, Little Wood, 68, 203
Savage River, 122, 123
Scaup, 165
Scaup, Lesser, 83, 90
Schist, 135, 136
Scoter, 83, 97, 181
Scott's Run Nature Preserve, 136
Screech-Owl, Eastern, 56, 67, 79, 109, 176,
    208
Sculpin, Mottled, 31
Seals: False Solomon's, 45; Harbor, 22; Harp,
    22; Hooded, 22; Many-flowered
    Solomon's, 45; Solomon's, 45
Seashore State Park, 171, 172, 193
Seatrout, Spotted, 31

Sedge: Tussock, 11; Umbrella, 11
Sedimentary, 3, 105, 126, 146, 152
Seneca, 89, 204
Seneca Creek, 203, 204, 205
Seneca Creek State Park, 141, 203
Seneca Rapids, 46, 132
Seneca Rocks, 103, 108, 109, 110
Seneca Shadows Campground, 109
Serpentine, 214
Serviceberry, 14
Shad, Atlantic, 31, 161
Shad Landing, 175
Shale, 10, 15, 35, 105, 125, 126, 135, 151
Shale Barrens, 10, 105, 126, 151
Sharpsburg, 128
Shavers Mountain Trail, 110, 111
Shelf, Sulphur, 16
Shells, 179, 190, 192
Shenandoah Mountain, 145
Shenandoah National Park, 24, 30, 72, 76,
    77, 91, 142, 144, 145, 146, 147, 148, 150
Shenandoah River, 128, 129, 150, 153
Shenandoah Valley, 6, 76, 128, 150, 151
Shepherdstown, 128
Shooting-Star, 106, 128
Shoveler, Northern, 84
Shrike, Loggerhead, 202
Sideling Hill, 125, 152
Signal Knob, 151
Sinks of Gandy, 109
Siskin, Pine, 90, 99, 100
Skimmer, Black, 189
Skinks: Broad-headed, 25; Coal, 25; Five-
    lined, 25, 54, 81, 149, 208; Ground, 25;
    Southeastern Five-lined, 25
Skippers: Clouded, 140; Common
    Checkered, 68; Fiery, 70; Gold-banded,
    104; Hobomok, 119; Leonard's, 215; Long-
    tailed, 140; Ocola, 140; Rare, 168; Salt
    Marsh, 184, 192; Silver Spotted, 33, 48
Skullcap, Showy, 149
Skunks: 133; Spotted, 20; Striped, 20
Skyline Drive, 146, 147
Slaughter Beach, 186
Sleepy Creek, 127, 152
Slider, Red-bellied, 24, 60, 71
Smith Center, 208

Smith Island, 176

Smoke Hole, 107, 108

Smoke Hole Caverns, 107

Snakes: Black Rat, 24, 60, 208; Brown, 24; Corn, 24, 71; Eastern Garter, 24, 54; Hognose, 53, 71, 191; Northern Water, 24, 53, 60, 71, 131; Redbelly, 24; Ring-necked, 24, 54; Rough Green, 24; Scarlet, 24; Smooth Earth, 24; Worm, 24

Snickers Gap, 81, 155

Snipe, Common, 131, 138

Snout, American, 201

Snowshoe Ski Resort, 114

Soldiers Delight Natural Environment Area, 214

Solomons Island, 170, 214

Sora, 213, 219

South Mountain, 154, 155

Sparrows: American Tree, 86, 90; Field, 217; Fox, 47; Grasshopper, 217; Henslow's, 56, 123; House, 99, 100; Ipswich, 98, 185; Saltmarsh Sharp-tailed, 91, 159, 169; Savannah, 98; Seaside, 56, 91, 169, 187; Swamp, 124, 131, 138; White-throated, 90, 93, 99

Sparrows Point, 163, 164

Spatterdock, 11, 13, 131

Speedwell, 96

Speicher, Darryl, 56

Spicebush, 77, 88, 96, 134

Spider, 38, 62

Spleenworts: Ebony, 15; Mountain, 15

Split Rock, 153

Spruce, Red, 9, 10, 13, 108, 110, 113, 116

Spruce Knob, 109, 145

Spruce Knob/Seneca Rocks National Recreation Area, 109

Spruce Mountain, 107, 109

Squirrels: Delmarva Fox, 21, 168; Fox, 21; Eastern Gray, 19, 21, 76, 88, 89, 91; Flying, 19, 21, 68, 79; Red, 21

Star, Blazing, 215

Starflower, 13

Stargrass, Yellow, 135

Starling, European, 99

Stars, 98, 99

Staunton, 145

Stick, Walking, 32

Stickleback, Fourspine, 31

Stinkpot, 24

Stint, Red-necked, 188

Stonecrop, 13

Stonefly, 32, 34, 49, 54, 74

Strasburg, 104

Stronghold, 202

Suffolk, 174

Sugarloaf Mountain, 6, 141, 201, 202

Sundew: 184; Round-leaved, 13

Sulphurs: Clouded, 48; Cloudless, 74, 82, 140; Pink-edged, 108

Sunfish, Pumpkinseed, 31

Sunflowers: Tickseed, 72; Woodland, 149

Susan, Black-eyed, 13

Susquehanna Flats, 162

Susquehanna National Wildlife Refuge, 162

Susquehanna River, 84, 158, 159, 161

Swallows: 153; Barn, 49; Eastern Tiger, 33, 48, 54, 68, 208; Northern Rough-winged, 49; Tree, 47, 49, 130, 165, 219

Swallowtails: Palamedes, 174; Spicebush, 54; Zebra, 48, 54

Swan Cove, 81

Swans: Trumpeter, 203; Tundra, 46, 55, 83, 87, 90, 165, 170, 202, 203

Sweetgum, 11, 12, 14, 138, 139, 168, 209, 214

Swift, Chimney, 134, 153

Switchgrass, 159

Sycamore, 10, 14, 108, 130, 173, 206, 217

Sycamore Landing, 130, 131, 132

Syncline, 152

T. Howard Duckett Reservoir, 211

Tamarack, 10, 123

Tanagers: Scarlet, 54, 123, 148, 201; Summer, 169, 176

Tangier Island, 176

Tangier Sound, 170

Teaberry, 66

Tea Creek Campground, 116

Teals: Blue-winged, 131; Green-winged, 84

Tearjacket Knob, 145

Tearthumb, Arrow-leaved, 11

Ted Harvey Wildlife Area, 186, 187, 188

Ten Mile Creek, 204

Terns: Black, 205; Common, 190; Least, 185; Royal, 171; Whiskered, 188; White-winged, 188

Terrapin, Diamondback, 24

Terrapin Trail Club, 143

Thomas, 168

Thompson Wildlife Area, 54, 148, 149

Thoroughfare Gap, 104

Thoroughwort, 82

Three Mile Island, 161

Three Ridges, 145

Three-square, American, 11

Three-square, Olney, 11, 159

Thrushes: Gray-cheeked, 52; Hermit, 77, 86, 90, 112, 114; Swainson's, 50, 114; Wood, 174

Ticks: 62; Deer, 38; Dog, 38

Titmouse, Tufted, 77, 90, 93, 99

Toads: American, 27, 66, 67, 78, 131, 204; Fowler's, 28

Tom's Cove, 81, 191

Toothwort, Cutleaf, 12, 44, 45

Towhee, Eastern, 184, 204, 219

Town Hill, 125

Townsend, George Alfred, 154

Tracks, 92, 93, 206

Transquaking River, 168

Tree, Cucumber, 14

Treefrogs: Gray, 28, 53; Green, 53; Pine Barrens, 184

Triadelphia Reservoir, 211

Trillium: 147; Large-flowered, 13, 54, 149; Painted, 123; Toadshade, 45, 133

Trout: 49, 54, 104, 116, 154; Brook, 31; Brown, 31, 210; Rainbow, 17, 31

Trout-lilies: 13, 45; White, 137

Tuliptree, 9, 10, 11, 14, 147, 153, 173, 206, 216

Turkey Run, 44, 136

Turkey, Wild, 91, 145, 151, 204

Turkey Point, 163

Turkey-tails, 16

Turtles: Box, 24, 166, 206; Eastern Mud, 24; Loggerhead Sea, 25; Painted, 24, 60, 71, 128, 133, 204, 211; Snapping, 24, 52, 71, 111; Spotted, 24, 174; Yellow-bellied, 173

Turtlehead, White, 13, 200

Tuscarora Mountain, 153

Tuscarora Trail, 150, 152, 153, 156

Twayblade: Large, 13; Southern, 174

Twinleaf, 45, 137

Twisted-stalk, Rose, 13

Tye River, 145

University of Maryland, 143, 167

Upper Marlboro, 212

U.S. Forest Service, 104, 107, 108, 110, 115

Valley and Ridge, 2, 3, 6, 8, 10, 11, 105, 107, 108, 125, 126, 129, 141–156, 161

Veach Gap, 151

Veery, 114, 145

Viburnum, 62, 209

Vienna, 221, 222

Violets: 44, 217; Birdsfoot, 106; Green, 54, 149

Violette's Lock, 46, 84, 132

Vireos: Red-eyed, 59, 148, 174, 214; Solitary, 49, 54, 148; Yellow-throated, 81, 201

Virginia, 15, 21, 22, 24, 30, 44, 54, 71, 81, 87, 90, 104, 119, 128, 129, 134, 137, 138, 139, 142, 143, 145, 147, 148, 149, 150, 153, 155, 158, 170, 174, 175, 176, 179, 191, 192, 193, 218, 220

Virginia Beach, 171, 193

Virginia Coast Reserve, 192

Voice of the Naturalist, 18

Voles: 78, 91; Meadow, 21; Southern Red-backed, 21

Waggoner's Gap, 81, 155

Wallrue, American, 15

Walnut, Black, 9, 10, 14, 130

Warblers: Black-and-white, 49; Black-throated Blue, 112, 116, 126, 145; Black-throated Green, 65, 111, 202; Black-burnian, 65, 111, 112; Blue-winged, 201; Canada, 65, 112; Cape May, 52; Cerulean, 135; Chestnut-sided, 65, 115, 126, 148, 153; Golden-winged, 108; Hooded, 54, 132, 135, 148, 174, 176; Kentucky, 54, 126, 135, 148, 175, 201; Magnolia, 109, 111, 112, 202; Mourning, 115; Nashville, 65, 117; Orange-crowned, 174; Palm, 49, 137, 219;

Warblers, (*continued*)
Pine, 49, 139, 181; Prairie, 126, 176, 184, 208, 222; Prothonotary, 49, 53, 174, 175, 204; Swainson's, 174; Tennessee, 52; Worm-eating, 132, 135, 148, 176, 201, 214; Yellow, 50, 64, 104, 131; Yellow-rumped (Myrtle), 75, 90, 93, 98, 109, 219; Yellow-throated, 49, 108

Wardensville, 104, 105

Warm Springs, 145

Washington and Old Dominion (W & OD) Trail, 221

Washington, D.C., 6, 7, 13, 28, 38, 44, 45, 47, 50, 54, 61, 63, 84, 90, 124, 125, 132, 133, 136, 137, 138, 154, 164, 170, 183, 195, 197, 205, 208, 209, 210, 212, 216, 218, 220

Washington, George, 171, 173

Washington Ditch, 173, 174

Washington Monument State Park, 155

Wasp, 32, 62, 70

Waterleaf, Virginia, 13

Waterthrushes: Louisiana, 49, 137, 204; Northern, 65, 117, 124

Waxwing, Cedar, 86, 90, 219

Waynesboro, 146

Weasel, Least, 20

Weasel, Long-tailed, 20

Webster Springs, 103

Weeds: Butterfly, 13, 127, 222; Joe-pye, 70; Rattlesnake, 135

Weevil, 32

Westmoreland State Park, 171

West Virginia, 13, 15, 30, 54, 65, 103–117, 119, 125, 127, 128, 129, 148, 153

Wheaton Regional Park, 196, 210

Whelk, 179, 190, 192

Whimbrel, 82

Whip-poor-will, 66, 109, 169, 184

White, Cabbage, 48

White Oak Canyon, 147

Widow, Black, 38

Wigeon, American, 46, 84

Wigeon, Eurasian, 176, 215

Wildcelery, 162

Wildfowl Trust of North America, 166

Wildrice, 11, 213

Willet, 82

Williamsport, 128

Williams River, 116

Willow, Black, 14

Wilmington, 163

Wintergreen: Common, 75; Spotted, 90

Woodbridge, 217, 218

Woodchuck (Groundhog), 20, 22, 86, 94, 208

Woodcock, American, 46, 55, 131

Woodend, 207

Woodpeckers: Downy, 90, 93, 133; Hairy, 90; Pileated, 53, 89, 133, 137; Red-bellied, 90, 93, 133; Red-headed, 90, 139, 203

Wood-pewee, Eastern, 59, 174, 214

Woodrat, Eastern, 21

Wrens: Carolina, 93; Marsh, 56, 159, 169; Winter, 86

Wye Island, 166

York, 161

York River, 158, 171

Yorktown, 171

Youghiogheny River, 123